CONCISE COLOR GUIDES

Modern Fighting Vehicles

General Editor
Bob Lewis

Longmeadow Press

This 1988 edition is published by
Longmeadow Press
201 High Ridge Road
Stamford CT 06904

©1988 Cheyprime Limited

All rights reserved. No part of this book may be reproduced or transmitted in any form or by any means, electronic or mechanical, including photocopying, recording or by any information storage or retrieval system, without written permission from the publisher.

ISBN 0 681 40433 7
Printed in Italy

0987654321

Contents

INTRODUCTION

All major nations have a strong force of main battle tanks and light tanks used for reconnaissance purposes. A main battle tank is a heavily armoured mobile gun platform capable of engaging other tanks and vehicles at ranges of up to 10km (6.2mls). Since 1944 tank development has reached technological levels then undreamt of. The first tank that can be considered modern is the British Centurion which was designed in 1944, and introduced into service in 1949. It proved its value in the Korean War, where it was able to defeat the Russian T-34/85. The Centurion is still in service with the Israeli Army nearly forty years after its introduction.

Tank development is constant, and because of the need and desire of opposing nations to maintain parity, if not superiority, of weaponry over their opponents, this need has led to the development of more sophisticated optical devices, improved firepower, better communications and more protection for the crews.

Historically the introduction of the Centurion into the Korean War in 1951 led the Russian designers to build the Josef Stalin 3 (JS-3) and the T-10, which in turn led to the development by Britain of the Conqueror in 1959. In 1966 the Conqueror was withdrawn from service and replaced with the smaller Chieftain, which had an improved firepower from a 120mm (4.7in) gun. The JS-3 with its 122mm (4.8in) gun had a longer range than the Centurion's 84mm (3.3in) gun.

Again Britain had to design a new gun, in this case the 105mm (4.1in), which was to be so successful in the Arab/Israeli conflicts.

During this time the United States, France and West Germany were pressing ahead with their own tank development programmes. In 1952 the United States introduced the M-48 into service and it remains in service to this day. A further improved version was introduced in 1960 as the M-60 and has gone through various modifications. In 1971 the USA instituted a research programme which has resulted in the production of M1 Abrams using a 120mm (4.7in) gun. The research contract was given to the Chrysler Corporation and to General Motors under the title of the XM1, with an instruction known as RAM-D (reliability, availability, maintainability and durability). The XM1 was to have a fire-control system equal to that envisaged by the United States' partners in NATO and to have the same protective armour as the Challenger.

One of the major problems faced by crews is storage of ammunition, equipment and tools. Faced with limited space, the crew have to ensure that equipment does not jam the turret traverse area and that they have enough food and water to survive if necessary in a nuclear environment enclosed in their tanks. Another problem faced by crews is fatigue; especially tiring are the night operations, where a man's effective observation limit is one hour, using modern sights. Comfort is also important, as a crew moving

Editor Bob Lewis commanding a Challenger tank, summer 1986

rapidly across an uneven terrain can emerge shaken and bruised, often with serious injury resulting from rapid cross-country movement. Research has resulted in the use of the Hydrogas suspension unit, which has dramatically reduced shaking and bruising. Research continues into suspension systems that can be retro-fitted to tanks already in service.

During the period when NATO was researching, developing and introducing its new generation of tanks, the Soviet Union was carrying out a similar programme. After the heavy and powerful JSIII and T-10 came the T-54/55, at 36 tonnes considerably lighter than its predecessors. Later came the T-62,

which was an onward development of the T-54/55 series. It is believed that over 40,000 T-62s were produced either in the USSR or in factories in the territories of its satellite nations. The T-62 weighed 40 tonnes and was armed with a 115mm (4.5in) gun. The early T-62 was relatively unsophisticated because of the Russian belief in massed armour attacks, along the lines of the German Blitzkrieg. However, later developments included laser range finders and infrared equipment. Russian tanks all have an ability that their NATO counterparts do not have – that of snorkelling or wading a river totally submerged. This ability exists because the Russians believe that, if war were to be joined, tanks would have to cross a great number of rivers often without the benefit of permanent or temporary bridges.

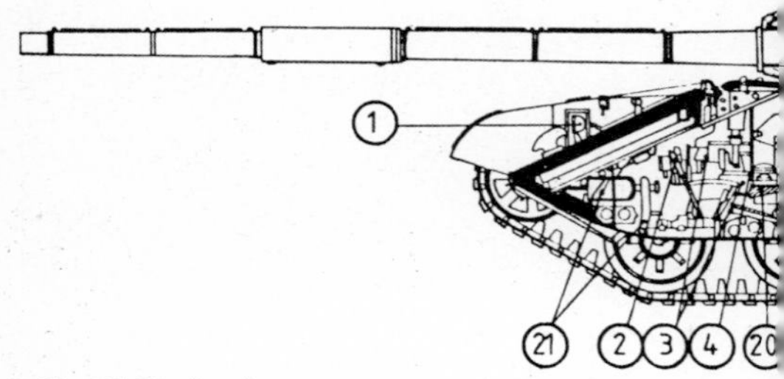

The T-62 also has two major variants, the flame thrower and the command vehicle which, apart from additional radio equipment, have the TNA navigation system incorporating a gyroscopic compass, latitude and longtitude correction devices, odometer, power converter and calculator.

A further Russian tank development is the T-64/72, which has a 125mm (4.9in) gun firing APFSDS, HEAT-FS and HE-FRAG FS from a smooth-bore gun, with an automatic loader, which enables the crew to be reduced to three men. From the T-64/T-72 has come a spectacular development which, while details are very limited, is believed to be the equal of the modern NATO-produced tanks. It is believed that

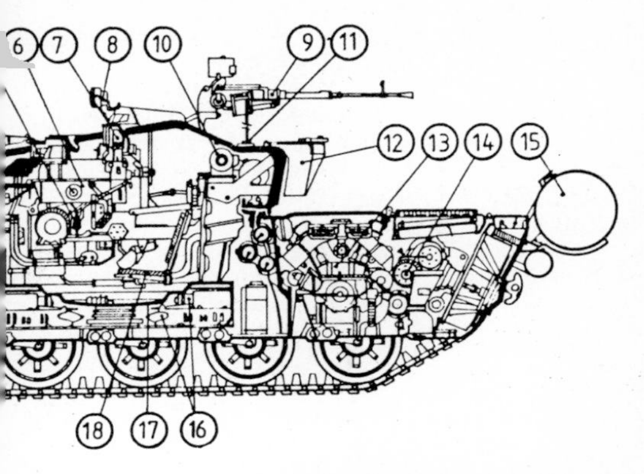

(**1**) *driver's FG 125 headlamp* (**2**) *steering tillers* (**3**) *NBC system* (**4**) *gear lever* (**5**) *gun elevation mechanism* (**6**) *TPD-2 gunner's sight* (**7**) *TPN 1-49-23 gunner's night sight* (**8**) *searchlight for use with TKN 3 observation device* (**9**) *12.7 mm anti-aircraft machine gun to rear* (**10**) *hoist for ammunition containers* (**11**) *antenna base* (**12**) *turret bin for deep fording equipment and cold rations* (**13**) *780 bhp diesel engine* (**14**) *gearbox* (**15**) *long-range fuel tank which can be jettisoned* (**16**) *charges and projectiles in containers on ammunition hoist platform* (**17**) *ammunition hoist platform* (**18**) *gunner's seat* (**19**) *NBC decontamination equipment* (**20**) *driver's adjustable seat*

this tank, known as the T-80, is based on the older T-64. With additional ceramic armour added to its turret, it incorporates a 125mm (4.9in) smooth-bore gun, which is now standard Soviet equipment. Research into new ammunition by the Russians has introduced a High Velocity Armour Pressing Fire Stabilized Discarding Sabot (HVAPFSDS), with further research into a Depleted Uranium (DU) round. In common with NATO tanks, the T-80 has been equipped with passive night-vision equipment, a laser rangefinder and a full-solution ballistic computer, as well as having full stabilization on the main armament.

At the same time as main battle tanks were being introduced by all nations, it was perceived that a lighter vehicle was needed for reconnaissance purposes, one that would not be heard or seen but with enough offensive weaponry to protect itself. Britain and France led the way, producing respectively the Saladin and the AMX-13 tank. The Saladin was a wheeled vehicle produced in 1958 with a 76mm (3in) cannon. The AMX-13 has a 90mm (3.5in) cannon, and weighs 15 tonnes. The French also introduced the Panhard 8-wheeled EBR armoured car; with a 90mm (3.5in) cannon and a driver's compartment at both ends with a crew of 4. Later requirements led to Britain introducing the Combat Vehicle Reconnaissance (Tracked) CVR(T), Scorpion with its 76mm (3in) gun; the Scimitar with a 30mm (1.18in) Rarden cannon; and the Combat Vehicle Reconnaissance

(Wheeled) (CVR(W)), Fox with a 30mm (1.18in) cannon. France also looked into further development, producing the successful AMX-10 series with a 105mm (4.13in) gun, the Renault VBC-90 with a 90mm (3.54in) gun, and the Panhard ERC-90 series with a variation of guns and turrets.

In the West, armoured cars have also subsequently been developed by the Brazilians, the Germans, the Israelis, the Italians and the United States – the US has versions currently in service.

The USSR has pressed ahead with its own reconnaissance vehicles, producing the BTR, BMP and PT76, as well as the BRDM, which is fitted with a variety of weapons. The Russians also have the PT76 light tank, which has a 76mm (3in) gun and a swimming capability. One of the mainstays of the Russian battalions is the T-62, which is also used in a reconnaissance role.

With the development of more sophisticated delivery systems capable of pinpoint accuracy, tank manufacturers have had to develop better protection, communications, fire power and mobility, as well as more automatic systems to protect their crews and ensure that the crews are able to carry out the functions they are tasked with.

Future developments will inevitably reduce the number of crew and produce a remotely controlled tank with fully automatic systems, and eventually an intelligent robot tank that is capable of making its own decisions and carrying out all the battle functions of the tank crews.

Abbreviations

A/A anti-aircraft
A/AG anti-aircraft gun
ACCV armoured cavalry cannon vehicle
ACRV armoured command and reconnaissance vehicle
ACV armoured cannon vehicle
AFARV armoured, forward-area, re-arm vehicle
AFD automatic feeding device
AFSV armoured fire-support vehicle
AFV armoured fighting vehicle
AGL above ground level
AGLS automatic gun-laying system
AGV assault gun vehicle
Ah ampére hour
AIFS advanced indirect fire system
AIFV armoured infantry fighting vehicle
AOS add-on stabilization
AP armour-piercing
APC armoured personnel carrier
APC-T armour-piercing capped tracer
APDS armour-piercing discarding sabot
APDS-T armour-piercing discarding sabot tracer
APERS anti-personnel
APERS-T anti-personnel tracer
APFSDS armour-piercing fin-stabilized discarding sabot
APHE armour-piercing high explosive
APIT armour-piercing incendiary tracer
APM anti-personnel mine
APT armour-piercing tracer
APU auxiliary power unit
ARETS armour remoted target system
ARMAD armoured and mechanized unit air defence
ARP anti-radiation projectile
ARSV armoured reconnaissance scout vehicle
ARV armoured recovery vehicle
ASV ammunition supply vehicle
A/T anti-tank
A/TG anti-tank gun
A/TGW anti-tank guided weapon
ATM anti-tank mine
ATTS automatic tank target system
ATWESS anti-tank weapons effects signature simulator
AVGP armoured vehicle, general purpose
AVLB armoured vehicle, launched bridge
AWESS automatic weapons effects signature simulator
BCC battery control centre
BD base detonating
BE base ejection
bhp brake horsepower
BITE built-in test equipment
BL blank
BL-T blank tracer
BOCV battery operations centre vehicle

BOFI Bofors optronic fire-control instrument
CAWS cannon artillery weapons systems
CET combat engineer tractor
CEV combat engineer vehicle
CF controlled fragmentation
CFV cavalry fighting vehicle
CGS crew gunnery simulator
CLGP cannon-launched guided projectile
COFT conduct-of-fire trainer
CP concrete-piercing
CPV command-post vehicle
CRT cathode ray tube
CSI computer-synthesized image
CSS computer sighting system
CVR(T) combat vehicle, reconnaissance (tracked)
CVR(W) combat vehicle, reconnaissance (wheeled)
CWR continuous-wave radar
DDU digital display unit
DHSS data handling sub-system
DIVADS division air-defense gun system
DOD Department of Defense
DSESTS direct-support electrical test system
DSO defence sales organization
DSWS division-support weapon system
EFCR equivalent full-charge rounds
EMG externally mounted gun
EOD explosive ordnance disposal
ER extended range
ERFB extended-range full-bore
ERGP extended-range guided projectile
ESRS electro slag refined steel
ERSC extended-range sub-calibre
ERV emergency rescue vehicle
ESPAWS enhanced self-propelled artillery weapons system
EW electronic warfare
FAAR forward-area alerting radar
FACE field-artillery computer equipment
FARS field-artillery rocket system
FAST fully automatic scoring/target system
FATMT field artillery turret maintenance trainer
FCCVS future close-combat vehicle system
FCE fire-control equipment
FCS fire-control system
FDC fire direction centre
FDCV fire-direction centre vehicle
FG field gun
FH field howitzer
FISTV fire-support team vehicle
FLIR forward-looking infrared
FMS Foreign Military Sales
FOO forward-observation officer
FROG free rocket over ground
FSCV fire-support combat

vehicle
FSED full-scale engineering development
FSMTC full-size moving tank target carrier
FV fighting vehicle
FV/GCE fighting-vehicle gun-control equipment
FVS fighting-vehicle system
FVSTMT fighting vehicle system turret maintenance trainer
GCE gun control equipment
GH gun/howitzer
GLC gun lay computer
GLLD ground laser locator designator
GP guided projectile
GPMG general purpose machine gun
GPO gun position officer
GPS gunners primary sight
GSR general staff requirement
GSRS general support rocket system
GST general staff target
GW guided weapon
HAWK homing-all-the-way-killer
HB heavy barrel
HC high capacity
HCHE high capacity high-explosive
HDTM heavy-duty tank target mechanism
HE high explosive
HEAT high-explosive antitank
HEAT-MP high-explosive anti-tank multi-purpose
HEAT-MP(P) high-explosive antitank multipurpose (practice)
HEAT-T high-explosive anti-tank tracer
HEAT-T-MP high-explosive anti-tank tracer multipurpose
HEDP high-explosive dual purpose
HEI high-explosive incendiary
HEIT high-explosive incendiary tracer
HELP howitzer extended life programme
HEP high-explosive plastic
HEP-T high-explosive practice tracer
HERA high-explosive rocket-assisted
HE-S high-explosive spotting
HESH high-explosive squash head
HE-T high-explosive tracer
HIMAG high-mobility agility test vehicle
How howitzer
hp horsepower
HPT high-pressure test
HSTV(L) high survivability test vehicle (lightweight)
HVAP high-velocity armour-piercing
HVAPDS-T high-velocity armour-piercing discarding sabot tracer
HVAPFSDS high-velocity armour-piercing fin stabilized discarding sabot
HVAP-T high-velocity armour-piercing tracer
HVSS horizontal volute spring suspension
HVTP-T high-velocity target

practice tracer
IAFV infantry armoured fighting vehicle
ICM improved convention munition
IFCS integrated fire-control system
IFV infantry fighting vehicle
IOC initial operation capability
IR infrared
IRBM inter range ballistic missile
ITV improved TOW vehicle
LAAG light anti-aircraft gun
LAD light aid detachment
LADS light air defence system
LAV light armoured vehicle
LAV light assault vehicle
LAW light anti-tank weapon
LDTM light-duty tank target mechanism
LED light emitting diode
LLLTV low-light level television
LMG light machine gun
LRU line replaceable unit
LTFCS laser tank fire-control system
LTD laser target designator
LVA landing vehicle, assault
LVT landing vehicle, tracked
LVTC landing vehicle, tracked command
LVTE landing vehicle, tracked engineer
LVTP landing vehicle, tracked personnel
LVTR landing vehicle, tracked recovery
LWT light weapon turret
MAC medium armoured car
MAV maintenance assist vehicle
MBT main battle tank
MDTM medium-duty tank target mechanism
MEV medical evacuation vehicle
MFC mortar fire controller
MG machine gun
MGB medium girder bridge
MICV mechanized infantry combat vehicle
MILES multiple integrated laser engagement system
MMS mast mounted sight
MoD Ministry of Defence
MPGS mobile protected gun system
MPWS mobile protected weapon system
MRS multiple rocket system
MRS muzzle reference system
MT mechanical time
MTG/WESS main tank gun/ weapons effects signature simulator
MTG/WESS main tank gun/ weapons effects simulator system
MTI moving target indication
MTSQ mechanical time and superquick
MULE modular universal laser equipment
MV muzzle velocity
NATO North Atlantic Treaty Organization
NBC nuclear, biological, chemical
OGS on-board gunnery

simulator
OP observation post
OT operational test
PAR pulse acquisition radar
PAT power-assisted traverse
PCB printed circuit board
PD point detonating
PE procurement executive
PFHE pre-fragmented high explosive
PGS platoon gunnery simulator
P How pack howitzer
PIVADS product improved Vulcan air defence system
PLARS position location and reporting system
PPI plan position indicator
PTO power take off
RAAMS remote anti-armour mine system
RAC Royal Armoured Corps
RAM-D reliability, availability, maintainability and durability
RAP rocket-assisted projectile
RATAC radar for field artillery fire
RCMAT radio-controlled miniature aerial target
RDF/LT rapid deployment force light tank
RDT & E research, development, test and evaluation
REME Royal Electrical and Mechanical Engineers
RHA Royal Horse Artillery
RISE reliability improved selected equipment
RMG ranging machine gun
ROBAT robatic counter-obstacle vehicle
ROC required operation characteristics
ROF Royal Ordnance Factory; rate of fire
ROR range only radar
RPV remotely-piloted vehicle
RSAF Royal Small Arms Factory
SADARM sense and destroy armour
SAL semi-active laser
SAM surface-to-air missile
SAPI semi-armour-piercing incendiary
SD self destruct
SFCS simplified fire-control system
SHORAD short range air defence system
SH/PRAC squash head practice
SLEP service life extension programme
SMG sub-machine gun
Smoke BE smoke base ejection
SP self-propelled
SPAAG self-propelled anti-aircraft gun
SPAG self-propelled assault gun
SPATG self-propelled anti-tank gun
SPAW self-propelled artillery weapon
SPG self-propelled gun
SPH self-propelled howitzer
SPL self-propelled launcher
SRG shell replenishment gear
STA shell transfer arm

STAFF smart target activated fire and forget
STATS stationary tank automatic target system
STUP spinning tubular projectile
SWAT special warfare armoured transporter
SWATT simulator for wire guided anti-tank tactical training
TACOM tank automotive command
TADS target acquisition and designation system
TALAFIT tank level aiming and firing trainer
TAS tracking adjunct system
TBAT TOW/Bushmaster armoured turret
TCT tank crew trainer
TCU tactical control unit
TD tank destroyer
TDCS tank driver command system
TES target engagement system
TGMTS tank gunnery training simulator
TGSM terminally guided sub-munition
TGTS tank gunnery tracking system
TI thermal imaging
TIRE tank infrared elbow
TIS thermal imaging system
TLS tank laser sight
TOGS thermal observation and gunnery system
TP target practice
TP-T target practice tracer
TTS tank thermal sight
TWGSS tank weapons gunnery simulation system
TWGSS tank weapons gunnery simulator system
ULC unit load container
USAADS US Army Air Defense school
VDM viscous damped mount
VDU visual display unit
VES vehicle engagement simulator
VHIS visual hit indicator system
WAPC wheeled armoured personnel carrier
WFSV wheeled fire support vehicle
WMRV wheeled maintenance and recovery vehicle
WP white phosphorus
WP-T white phosphorus tracer

AMX-32
MAIN BATTLE TANK

Country of Origin: France
Crew: 4

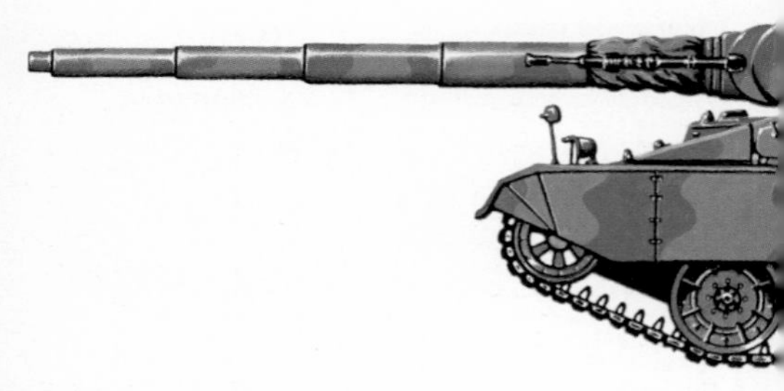

cooled supercharged multifuel, developing 70hp at 2,400rpm
Performance: Speed 65km/h (40mph); range 530km (329mls) (road)
Armament: 1 × 105mm (4.16in) gun (or 1 × 120mm (4.75in) gun); 1 × 20mm (1in) coaxial cannon; 1 × 7.62mm (0.3in) machine gun
Dimensions: Length (gun forward), 105mm: 9.45m (30ft 11in); 120mm: 9.85m (32ft 3in); length (hull) 6.59m (21ft 7in); width overall 3.25m (10ft 7in); height 2.96m (9ft 8in) (turret roof 2.29m (7ft 6in))
Ground Clearance: 0.45m (1ft 5in)
Ground Pressure: 0.90kg/cm^2 (12.87psi)
History: The first prototype AMX-32, armed with the

105mm (4.16in) CN-105-F1 gun, was completed in 1979, but subsequently the turret was completely redesigned and the 120mm (4.75in) smooth-bore gun can be fitted as an option. Mounted to the left of the turret is a 20mm (0.79in) M693 (F2) cannon which, although coaxially mounted, can be elevated independently to +40° for use against low-flying aircraft or helicopters. Fire control is effected by an integrated COTAC system originally developed for the AMX-10RC, whilst the commander's all-round visibility is enhanced by the new TOP7VS cupola capable of counter-rotating a full 360°. The turret-mounted 7.62mm (0.3in) machine gun rotates itself around the cupola and can be aimed and fired by the commander.

AMX-30 MAIN BATTLE TANK

Country of Origin: France

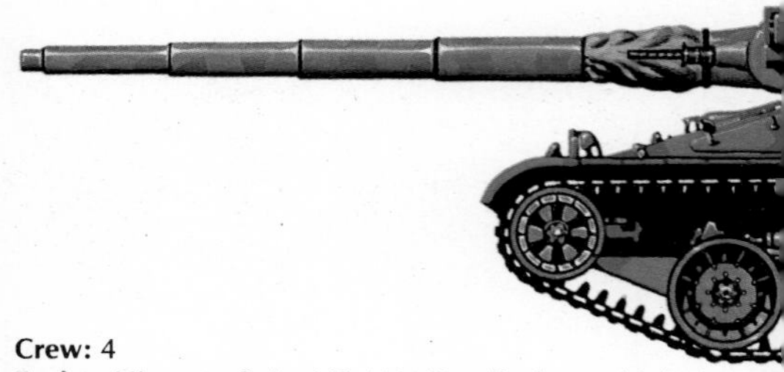

Crew: 4

Engine: Hispano-Suiza HS-110, 12-cylinder multi-fuel, water-cooled engine developing 720hp at 2,400rpm

Performance: Speed 65km/h (40mph); range 600km (373mls) road; fuel 970lit (213gal)

Armament: 1 × 105mm (4.13in) gun, elevation +20°, depression −8°, (50 rounds carried); 1 × 12.7mm (0.5in) MG or 20mm (0.79in) cannon coaxial with main armament, it can be elevated to +40° (1,050 rounds carried); 1 × 7.62mm (0.3in) MG on commander's cupola (2,050 rounds carried); 2 × 2 smoke grenade launchers

Dimensions: Length 9.48m (31ft 1in) with gun forward; 6.59m (21ft 7in) hull only; width 3.10m (10ft 2in); height 2.86m (9ft 4.5in) to top of cupola

Weight: 36,000kg (79,350lb) loaded

Ground Pressure: 0.77kg/ cm^2 (10.95lb/sq in)

History: The AMX-30 should have been an international development between France , Italy and Germany and an agreement to build a common tank was reached in 1956. Italy and Germany later pulled out leaving France to go it alone. The first prototype was ready in 1960, a second in 1961, and pre-production models developed between 1963 and 1965. The tank entered production in 1966 and deliveries commensed in mid 1967. The AMX-30 is fitted with infra-red driving and fighting equipment whilst the AMX-30B2 later production variant is also fitted with laser rangefinder and automatic COTAC integrated fire control system. Other variants are the AMX-30 Export Model (without infra-red equipment); the AMX-30S fitted with sandshields for use in hot climates; and the AMX-30 Shatine.

LEOPARD 2
MAIN BATTLE TANK

Country of Origin: West Germany
Crew: 4

Engine: MTU MB 873 Ka 501 4-stroke, 12-cylinder multifuel, exhaust turbocharged, liquid-cooled, developing 1500hp at 2,600rpm
Performance: Speed 72km/h (44.72mph) (forward), 31km/h (19.25mph) (reverse); range 550km (341.5mls)
Armament: 1 × 120mm (4.75in) tank gun; 1 × 7.62mm (0.3in) coaxial machine gun; 1 × 7.62mm (0.3in) A/A machine gun
Dimensions: Length (gun forward) 9.688m (31ft 8in); length (hull) 7.722m (25ft 3in); width overall (including skirts) 3.7m (12ft 1in); height (turret top) 2.46m (8ft); ground clearance (front) 0.53m (1ft 8in), (rear)

0.48m (1ft 6in)
Ground Pressure: 0.81kg/cm² (11.58psi)
History: Representing the latest in West German technology, the Leopard 2 combines maximum protection with high mobility and enhanced firepower. Main armament consists of the Rheinmetall 120mm (4.72in) smooth-bore gun with a dop-type breech block, fume extractor and thermal sleeve. Two types of ammunition are fired, both fin-stabilized: the APFSDS-T, with an effective range well in excess of 2,000m (1.2mls); and the HEAT-MP-T, with a proven effectiveness against both soft and hard targets.

LEOPARD I
MAIN BATTLE TANK

Country of Origin: West Germany
Crew: 4

Engine: MTU MB 838 Ca. M500, 10-cylinder multifuel engine, developing 830hp at 2,200rpm
Performance: Speed 65km/h (40mph) (road); range 500km (372mls) (road), 450km (280mls) (cross-country); fuel 955 litres (252gal)
Armament: 1 × 105mm (4.13in) L7A3 gun, elevation +20°, depression – 9° (stabilized) (60 rounds carried); 2 × 7.62mm (0.3in) MGs – coaxial and A/A (5,500 rounds); 4 smoke dischargers either side of turret
Armour: 10mm-70mm (0.4in-2.75in) (estimate)
Dimensions: Length 9.54m (31ft 3in) (gun forward), 7.09m (23ft 3in) (hull only); width 3.25m (10ft 8in), (3.40m (11ft 2in) with skirts); height 2.61m (8ft 6in)

(commander's periscope)

Weight: 40,000kg (88,185lb) (loaded), 38,700kg (85,319lb) (empty)

Ground Pressure: 0.90kg/cm² (12.87psi)

History: The first production Leopard I MBT was delivered in September 1965 and built by Krauss-Maffei of Munich. The original intention in the mid 50s was for Italy and France to work with Germany on the development of a standard tank, but disagreement led to France building their AMX-30 and Italy the M60A1, although Italy did purchase Leopards in 1970. Twenty-six prototypes were produced before mainline production commenced in July 1963.

TAM
MAIN BATTLE TANK

Country of Origin: West Germany
Crew: 4
Engine: Supercharged MTU V-6 diesel, developing

710hp at 2,200rpm
Performance: Speed 75km/h (46mph); range 550km (341mls); 900km (559mls) with long-range fuel tanks; fuel 650 litres (172gal)
Armament: 1 × 105mm (4.13in) gun, elevation +18°, depression –7° (50 rounds carried); 2 × 7.62mm (0.3in) MGs – coaxial and A/A (6,000 rounds); 4 smoke dischargers either side of turret
Dimensions: Length 8.23m (27ft) (gun forward), 6.775m (22ft 2in) (hull); width 3.25m (10ft 8in); height 2.42m (8ft) (turret top)

Weight: 30,500kg (6,7240 lb)
Ground Pressure: 0.79kg cm^2 (11.30psi)

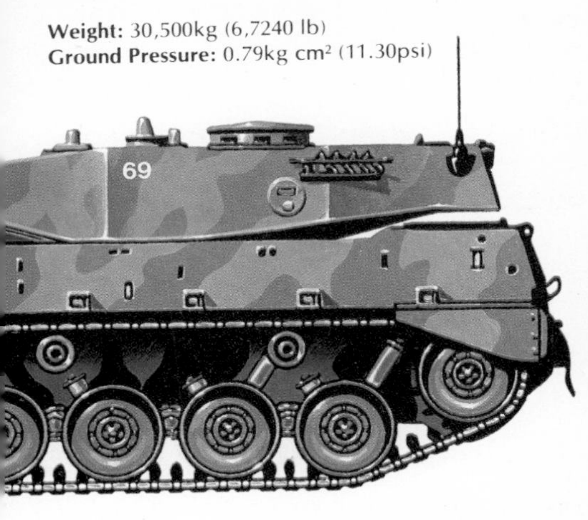

History: Designed specifically for the Argentinian army, the Tanque Argentino Mediado (TAM) tank was originally developed and built by Thyssen Henschel in West Germany, although it is now produced in Argentina with Thyssen Henschel still making many of the component parts. The first prototypes were produced in 1976 and were based on the Marder MICV chassis, but with a more powerful engine. Variants include an infantry fighting vehicle with a two-gun turret, and an improved model (the TAM-4) is under development.

MERKAVA
MAIN BATTLE TANK

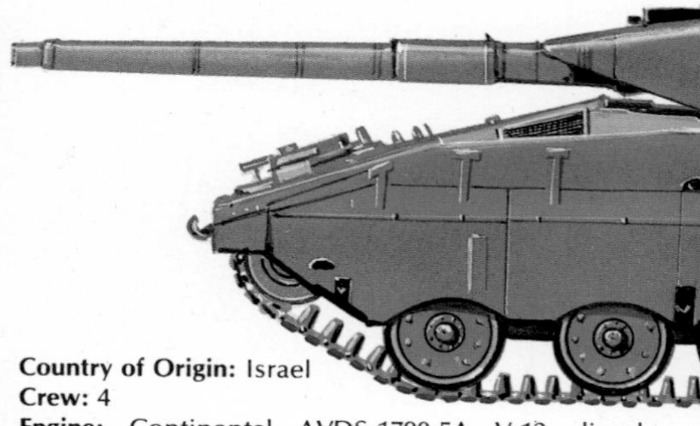

Country of Origin: Israel
Crew: 4
Engine: Continental AVDS-1790-5A V-12 diesel, developing 900hp at 2,400rpm
Performance: Speed 46km/h (28.5mph); range 500km (310mls); fuel 1,200 litres (317gal) (estimate)
Armament: 105mm (4.13in) gun, elevation +20°, depression –8° (65+ rounds carried); smoke dischargers and 60mm (2.36in) roof-mounted mortar
Dimensions: Length 8.63m (28ft 4in) (gun forward), 7.45m (24ft 5in) (hull); width 3.7m (12ft 7in); height 2.64m (8ft 8in) (turret roof)
Weight: 56,000kg (12,345lb)

History: Production models of the Merkava MBT became available in 1979, following two years after the first prototypes. Night-vision equipment is standard, and the fire-control system includes a laser rangefinder. Engine and transmission are positioned at the front, whilst doors at the rear provide a facility for the quick reloading of ammunition. The relatively small turret is also placed towards the rear. Variants include the Mk2 with hydropneumatic suspension, and most recently the Mk3 incorporating a larger-calibre gun.

UPGRADED CENTURION
MAIN BATTLE TANK

Country of Origin: Israel
Crew: 4
Engine: Teledyne Continental AVDS-1790-2A diesel
Performance: Speed 43km/h (26.7mph) (road); range 380km (226mls) (road); fuel 1,037 litres (274gal)

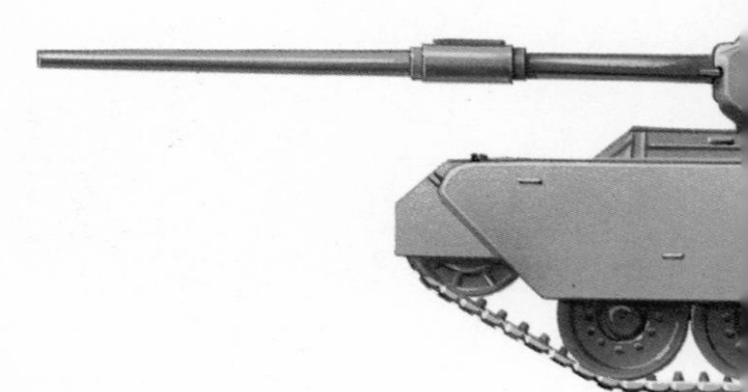

Armament: 1 × 105mm (4.13in) L7-series rifled tank gun; 1 × 12.7mm (0.5in) RMG (600 rounds); 2 × 7.62 mm (0.13in) MGs – coaxial and A/A (4,750 rounds); 2 × 6-barrelled smoke dischargers
Armour: 17mm-152mm (0.67in-6in)
Dimensions: Length 9.854m (32ft 4in) (gun forward), 7.823m (25ft 8in) (hull); width 3.39m (14ft 1in); height 3.009m (9ft 10in) (turret roof)
Weight: 51,820kg (loaded) (114,243lb)
Ground Pressure: 0.95kg/cm^2 (13.58psi)

History: The original Centurion supplied to the Israeli Army in 1959 was powered by the standard Meteor petrol engine, which offered only a limited range and a low power-to-weight ratio. It was replaced in 1970 by the Teledyne diesel engine, along with an Allison CD-850-6 automatic gearbox to replace the original Merritt-Brown Z51R manual. During the refits the rear of the hull was enlarged, and elevated top decks were added to accommodate air vents. Finally, new fire-extinguishing and electrical systems were added.

OF-40
MAIN BATTLE TANK

Country of Origin: Italy
Crew: 4
Engine: 90° V-10 supercharged multi fuel developing 830hp at 2,200rpm

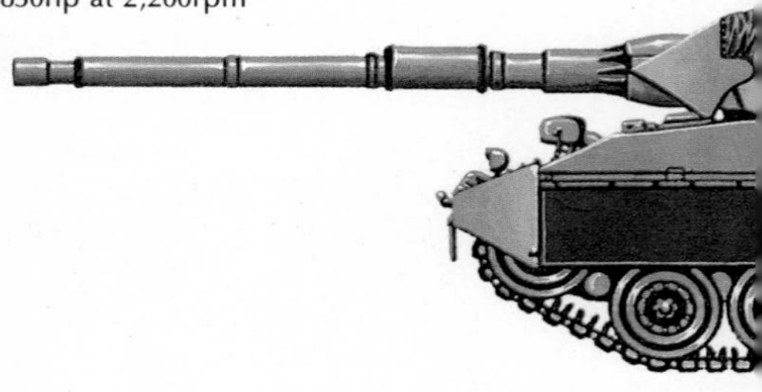

Performance: Speed 60km/h (37.25 mph); range 600km (372mls) (road); fuel 1,000 litres (264gal)
Armament: 1 × 105mm (4.13in) gun, elevation +20°, depression –9° (61 rounds carried); 2 × 7.62mm (0.13in) MGs – coaxial and A/A (5,500 rounds); 4 smoke dischargers either side of turret
Dimensions: Length 9.222m (30ft 3in) (gun forward), 6.893m (22ft 7in) (hull); width 3.51m (11ft 6in) (with skirts); height 2.68m (8ft 11in) (commander's sight)
Weight: 43,000kg (94,798lb) (loaded), 40,000kg (88,185lb) (empty)
Ground Pressure: 0.86 kg/cm^2 (12.3psi)

History: The first production tanks became available in 1981, only a year after the first prototype. Fiat developed the engine whilst OTO-Melara were responsible for the hull, chassis and armament. The OF-40 main battle tank is equipped to fire standard NATO ammunition such as APDS, HEAT and HESH. A laser rangefinder is standard, together with a roof-mounted SFIM 580-B sight. An over-pressure NBC system is fitted, as is an automatic fire-extinguishing system in the engine compartment. Variants include OF-40s fitted with alternative weapon systems, including an anti-aircraft turret.

TYPE 74
MAIN BATTLE TANK
Country of Origin: Japan

Crew: 4
Engine: Mitsubishi 10ZF Type 21 WT, 10-cylinder diesel, air-cooled, developing 750hp at 2,200rpm
Performance: Speed 53km/h (33mph) (road); range 300km (186mls), fuel 950 litres (251gal)
Armament: 1 × 105mm (4.13in) gun (51 rounds carried; 1 × 7.62mm (0.31in) coaxial MG (4,500 rounds); 1 × 12.7mm (0.5in) M2 A/A MG (660 rounds); 1 × 3 smoke dischargers either side of turret
Dimensions: Length 9.41m (30ft 10in) (gun forward), 6.7m (21ft 11in) (hull); width 3.18m (10ft 2in); height 2.67m (8ft 11in) (normal)

Weight: 38,000kg (83,775lb) (loaded), 36,300kg (80,028lb) (empty)
Ground Pressure: 0.86kg/cm^2 (12.3psi)
History: Production of the Type 74 MBT commenced in 1973 at the Mitsubishi Heavy Industries Works, following ten years of testing and prototype development. Designed to replace the Type 61, production vehicles became available in September 1975 fitted with adjustable hydropneumatic suspension which allows the height of the tank to be altered. Infrared driving lights are fitted, plus a laser rangefinder, an NBC system, and gun stabilizers.

TYPE 61
MAIN BATTLE TANK

Country of Origin: Japan
Crew: 4
Engine: Mitsubishi Type 12 HM 21 WT, V-12, turbo-

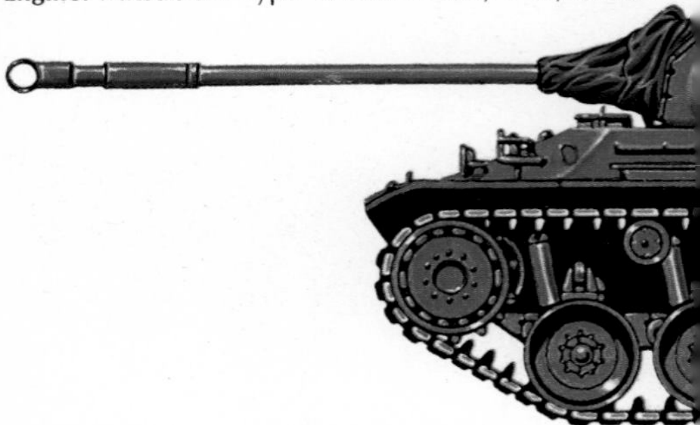

charged air-cooled diesel, developing 600hp at 2,100rpm
Performance: Speed 45km/h (28mph) (road); range 200km (124mls) (road)
Armament: 1 × 90mm (3.54in) Type 61 gun; 1 × 7.62mm (0.31in) M1919A4 coaxial MG; 1 × 12.7mm (0.5in) M2 A/A MG
Armour: 64mm (2.52in)
Dimensions: Length 8.19m (26ft 10in) (gun forward), 6.30m (20ft 8in) (hull); width 2.95m (9ft 8in)
Weight: 35,000kg (77,162lb) loaded

Ground Pressure: 0.95kg/cm^2 (13.59psi)
History: The first Japanese postwar tank, the Type 61 was built under the direction of the Ground Armaments Directorate at the headquarters of the Japanese Self-Defence Forces. Design work started in 1954 and prototypes were produced in 1957, with production vehicles being available in 1962. A total of about 560 Type 61 MBTs were built with standard fitments, including infrared driving and fighting equipment.

STRIDSVAGN 103B (S TYPE)
MAIN BATTLE TANK

Country of Origin: Sweden
Crew: 3

Engine: Rolls-Royce K60 multi fuel, developing 240bhp at 3650rpm and Boeing 553 gas-turbine developing 490shp at 38,000rpm
Performance: Speed 50km/h (31 mph) land; 6km/h (3.7 mph) water; range 390km (192 mls)
Armament: 1 × 105mm (4.16in) L7A1 L/62 gun; 2 × 7.62mm (0.3in) coaxial machine guns; 1 × 7.62mm (0.3in) A/A machine gun
Dimensions: Length (gun forward) 9.8m (32ft 1in), length (hull) 8.4m (27ft 6in); width overall 3.6m (11ft 9in); height 2.14m (7ft) (commander's cupola), 2.5m (8ft 2in) (with commander's machine gun); ground clearance 0.5m (1ft 7in) (centre of hull), 0.4m (1ft 3in) (sides of hull)
Ground Pressure: 0.9kg/cm² (12.87psi)
History: The S-Tank is unique in that it has no turret,

the gun being located in the hull. The gun itself is designated the L7, being in essence a stretched version of the conventional British L7-series gun. The gun is fed externally from a magazine with a 50-round capacity. The automatic loader enables a fast rate of fire of 15 rounds per minute, whilst the magazine itself can be reloaded via two hatches at the rear in approximately ten minutes. A mixed load of ammunition can be carried and can include APDS with an effective range of over 2,000m (1.26mls), HE with a range of 5,000m (3.15mls), and smoke. Two fixed 7.62mm (0.3in) coaxial KSP58 machine guns are mounted on the left of the turret and can be fired. These guns can only be loaded externally, one of the crew having to leave the vehicle to do so.

PZ 68
MAIN BATTLE TANK

Country of Origin: Switzerland
Crew: 4

Engine: MTU MB 837 V-8 diesel, developing 660bhp at 2,200rpm
Performance: Speed 55km/h (34mph) (road); range 350km (217mls) (road); fuel 710 litres (187 gal)
Armament: 1 × 105mm (4.13in) gun, elevation +21°, depression -10° (56 rounds carried); 2 × 7.5mm (0.39in) MGs – coaxial and A/A (5,200 rounds); 6 smoke dischargers
Armour: 60mm (2.36in) (maximum)
Dimensions: Length 9.49m (31ft 1in) (gun forward), 6.98m (22ft 11in) (hull); width 3.14m (10ft 4in); height 2.75m (9ft 0in) (cupola)
Weight: 39,700kg (87,532lb) (loaded), 38,700kg (85,319lb) (empty)
Ground Pressure: 0.86kg/cm^2 (12.3psi)

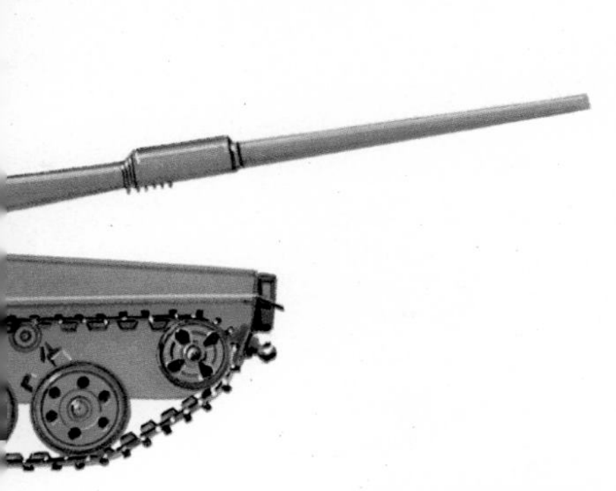

History: The first production Pz (Panzer) 68s were delivered in 1971, being a development of the Pz 61 MBT which in 1961 had become Switzerland's first home-produced MBT. The Pz 68s have only been supplied to the Swiss Army and differ from the earlier Pz 61 mainly in the replacement of the 20mm (0.79in) coaxial cannon with a standard 7.5mm (0.3in) MG, a stabilization system for the main 105mm (4.13in) gun (enabling the tank to engage targets accurately whilst moving across country), wider tracks, and a more powerful engine with improved transmission. The Mk2 appeared in 1974, the Mk3 in 1978, and the Mk4 in 1981; there is very little difference between these except for a thermal sleeve for main armament and a larger turret.

T-10M
HEAVY TANK

Country of Origin: USSR
Crew: 4

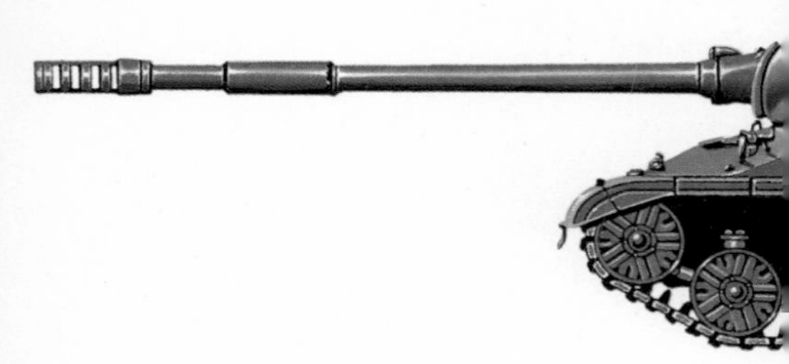

Engine: V-2-IS, 12-cylinder, water-cooled diesel, developing 700hp at 2,000rpm
Performance: Speed 42km/h (26mph) (road); range 250km (155mls) (road); fuel 900 litres (238 gal)
Armament: 1 × 122mm (4.80in) gun, elevation +17°, depression -3° (30 rounds carried); 2 × 14.5mm (0.57in) KPV MGS – coaxial and A/A (1,000 rounds)
Armour: 230mm (9.05in) (maximum)
Dimensions: Length 10.6m (34ft 9in) (inc gun), 7.04m (23ft 1in) (exc gun); width 3.566m (11ft 8in); height 2.43m (8ft 0in) (w/o A/A MG)
Weight: 52,000kg (114,640lb)
Ground Pressure: 0.78 kg/cm^2 (11.15psi)

History: The T-10 entered service in 1955 and is a direct descendant of the JS series of heavy tanks. The T-10 has seven road wheels, a larger turret than the JS-3, and cut-off rear hull-plate corners. The T-10 took part in the 1968 invasion of Czechoslovakia, but only a very few are still in service with the Soviet Army today. The T-10 was followed by the T-10M which had a number of improvements. The 122mm (4.8in) gun was fitted with a multi-baffle muzzle brake, and was stabilized in both horizontal and vertical planes. The T-10M can also be fitted with a snorkel and has an infrared searchlight; an overpressure NBC system has also been fitted.

JS-3
HEAVY TANK

Country of Origin: USSR
Crew: 4

Engine: V-2-IS, V-12 diesel, water-cooled, developing 520hp at 2,000rpm
Performance: Speed 37km/h (23mph) (road); range 150km (93mls) (road); fuel 520 litres (137 gal)
Armament: 1 × 122mm (4.80in) M1943, elevation +20°, depression -3° (28 rounds carried); 1 × 7.62mm (0.30in) DTM coaxial MG (1,500 rounds); 1 × 12.7mm (0.50in) D Sh KM AA MG (250 rounds)
Armour: 20mm-200mm (0.79in – 7.87in)
Dimensions: Length 9.725m (3ft 11in) (inc gun), 6.77m (22ft 2in) (exc gun); width 3.07m (10ft 1in); height 2.44m (8ft 0in) (w/o A/A MG)
Weight: 45,800kg (100,971lb) (loaded)

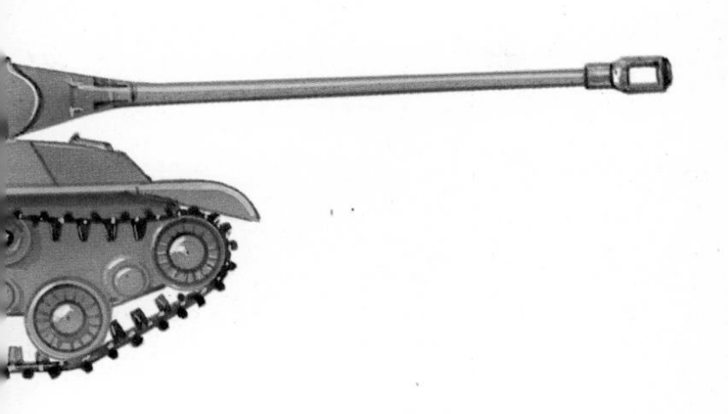

Ground Pressure: 0.83kg/cm^2 (11.87psi)
Ground Clearance: 0.46mm (1.81in)
History: The JS-3 entered production in 1945 having the same engine, transmission and armament as its predecessor, the JS-2, although it featured a redesigned hull and larger turret. Subsequent developments led to the JS-4, JS-5, JS-6, JS-7, JS-8, JS-9 and JS-10 (which was redesignated T-10). Production was at the Leningrad Kirov Plant, which manufactured 3750 JS heavy tanks in a 20-month period from October 1943. Now withdrawn from front-line service, JS-3s are used for training purposes, and many are still held in reserve.

T-80
MAIN BATTLE TANK

Country of Origin: USSR
Crew: 3

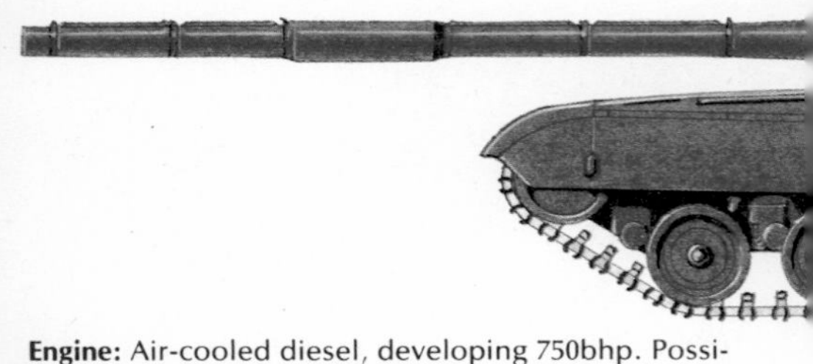

Engine: Air-cooled diesel, developing 750bhp. Possibly an update of the T-64 engine.
Performance: Speed 70km/h (43.5mph); range 650km (403.7mls) (with long distance fuel tanks)
Armament: 1 × 122mm (4.83in); 1 × 7.62mm (0.3in) co-axial machine gun; 1 × 12.7mm (0.5in) A/A gun (or 14.5mm (0.57in) A/A machine gun)
Dimensions: Length (full) 7.00m (22ft 11in); width overall 3.5m (11ft 5in); height 2.25m (7ft 4in); ground clearance variable
Weight: 48,500kg (106,700lb)
History: Although now operational in the Group of

Soviet Forces in Germany, little information is available about the T-80, and even the above statistics are provisional. The T-80 appears to incorporate the best

of the T-64 and the T-72. Suspension is similar to that of the T-64, with six small road wheels, idlers at the front, and back and return rollers at the rear. Track protection is afforded by light steel skirts, not unlike those found on the Chieftain, whilst the glacis plate is well-sloped with a deep V splash board. Main armament is provided by one 125mm (4.92in) gun similar to that of the T-64/T-72, although it is possible that the ammunition has been improved for enhanced penetration. Some sources suggest T-80 can fire guided-missile projectiles.

T-64
MAIN BATTLE TANK

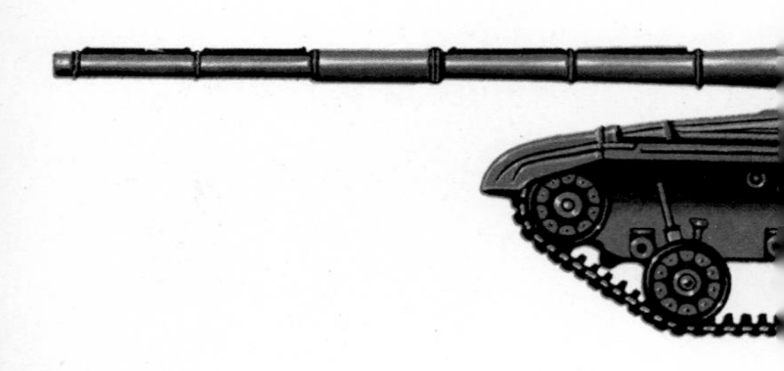

Country of Origin: USSR
Crew: 3
Engine: 5-cylinder opposed piston, diesel, liquid-cooled, developing 700/750hp at 3000rpm
Performance: Speed 70km/h (43.5mph); range 450km (279mls) (road) (700km with long range fuel tanks fitted).
Armament: 1 × 125mm (4.92in) gun; 1 × 7.62mm (0.3in) coaxial gun; 1 × 12.7mm (0,5in) A/A gun
Dimensions: Length (gun forward) 9.1m (29ft 10in); length (full) 6.4m (21ft 0in); width overall 4.64m (15ft 2in) (with skirts), 3.38m (11ft 1in) (without skirts);

height 2.4m (7ft 10in); ground clearance 0.377mm (14.8in)
Ground Pressure: 1.09kg/cm² (15.58psi)
History: The T-64 was first seen in 1970 and is in service solely with the Soviet Army, both domestically and with the Group of Soviet Forces in Germany. The long 125mm (4.83in) smooth-bore gun fires Armour Piercing Fin-Stabilized Discarding Sabot (APFSDS) rounds to an estimated effective range of 2,000m (1.2mls). The 12.7mm (0.5in) A/A gun mounted on the commander's cupola can be fired automatically from inside the turret.

T-72
MAIN BATTLE TANK
Country of Origin: USSR

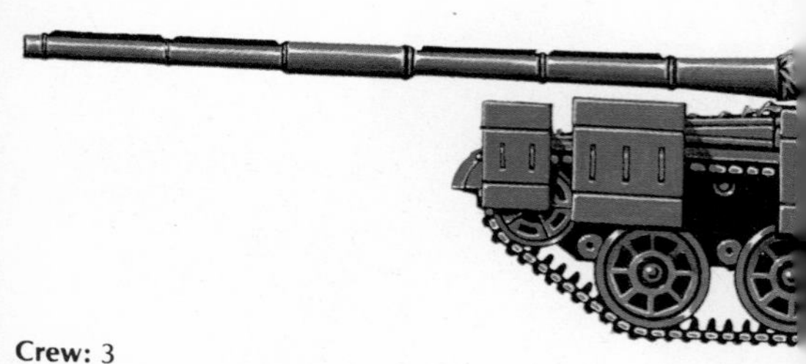

Crew: 3
Engine: V-12 Diesel (V-46), developing 780hp at 3,000rpm
Performance: Speed 60km/h (37mph); range 480km (298mls) (road), 700km (434.8mls) with long range fuel tanks provided)
Armament: 1 × 125mm (4.92in) gun; 1 × 7.62mm (0.3in) coaxial gun; 1 × 12.7mm (0.5in) A/A gun
Dimensions: Length (gun forward) 9.2m (30ft 2in) length (full) 6.9m (22ft 7in); width overall 4.75m (15ft 7in) (with skirts), 3.6m (11ft 9in) (without skirts); height 2.37m (7ft 9in); ground clearance 0.47m (1ft 6.5in)
Ground Pressure: 0.83kg/cm² (11.87psi)
Weight: 41,000kg (25,500lb) loaded

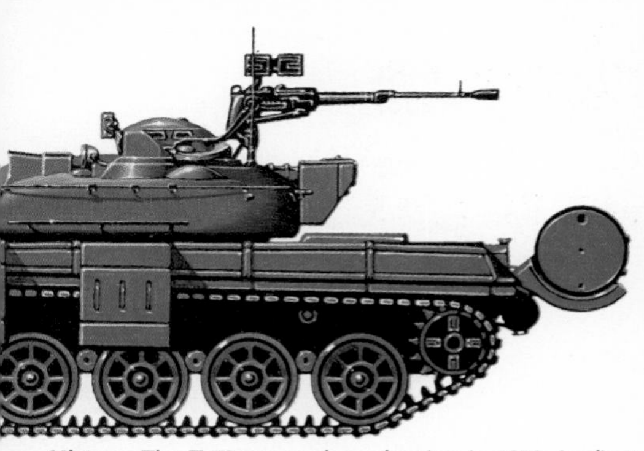

History: The T-72 entered production in 1972. Its first public sighting was at the Moscow military parade of November 1977. Since then production has continued at six factories at the rate of 2,000 tanks per annum. Main armament is provided by the 125mm (4.92in) smooth-bore gun fitted with a light-alloy thermal sleeve and bore evacuator. Three types of separate loading ammunition are fired: APFSDS with a maximum range of 2,100m (1.3mls), HEAT-FRAG with a maximum direct fire range of 4,000m (2.5mls), and smoke with a maximum indirect range of 9,400m (5.8mls). Firing on the move is possible, whilst the automatic loading system affords a high rate of fire of up to eight rounds per minute, although reliability of the system is in doubt.

T-62
MAIN BATTLE TANK

Country of Origin: USSR
Crew: 4
Engine: Model V-12 water-cooled diesel engine,

developing 580hp at 2,000rpm
Performance: Speed 45.5km/h (28.26mph) (road); range 450km (279.5mls) (road); fuel 1,360 litres (299.56gal) (total)
Armament: 1 × 115mm (4.55in) (U-5TS), elevation +17°, depression -4° (40 rounds carried); 1 × 7.62mm (0.3in) PKT coaxial MG (2,500 rounds)
Armour: 20mm-170mm – (0.8in)-(6.43in)
Dimensions: Length 9.335m (30ft 7in) (inc gun), 6.63m (21ft 8in) (exc gun); width 3.3m (10ft 9in); height 2.395m (7ft 10in); ground clearance 0.425m (1ft 4in)
Weight: 40,000kg (88,000lb) (loaded)
Ground Pressure: 0.83kg/cm² (11.87psi)
History: First seen in 1965 at the May Day Parade, the

T-62 Main Battle Tank had then been in production for four years and was a development of the T-54. The T-62, in addition to being a longer tank, had a new turret and new 115mm (4.52in) smooth-bore gun, which was fitted with a fume extractor and despatched the empty cartridge cases through the rear of the turret. The gun is capable of firing APFSDS with a muzzle velocity of 1,680m/sec (5,600ft/sec), HEAT with a muzzle velocity of 1,000m/sec (3,333ft/sec), and two forms of HE fragmentation. Variants include the T-62A, which has a 12.7mm (0.5in) A/A MG fitted to the loader's hatch, and the command model T-62K. Infrared night-driving and night-fighting equipment is standard.

T-55
MAIN BATTLE TANK
Country of Origin: USSR
Crew: 4

Engine: Model V-55 V-12 water-cooled diesel, developing 580hp at 2000rpm
Performance: Speed 50km/h (31.1mph); range 500km (310mls)
Armament: 1 × 100mm (36in); 1 × 7.62mm (0.3in) coaxial gun; 1 × 7.62mm (0.3in) bow machine gun; 1 × 12.7mm (0.5in) A/A machine gun (where fitted)
Armour: 20-203mm (0.79-8in)
Dimensions: Length (gun forward) 9.02m (29ft 6in); length (full) 6.45m (21ft 2in); width overall 3.27m (10ft 8in); height 2.4m (7ft 9in); ground clearance .425m (1ft 4in)
Ground Pressure: 0.8kg/cm^2 (11.44psi)
History: The T-54 first entered service in the Soviet Union in 1949 and was followed by the T-55 in 1958, in

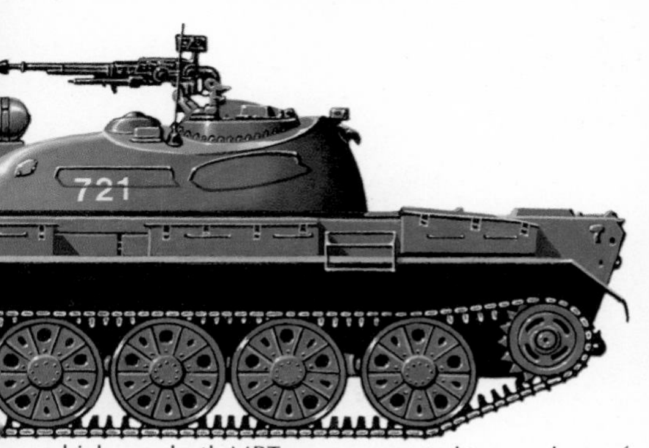

which year both MBTs were exported to members of the Warsaw Pact. Since then, over 40,000 T-54s and T-55s have been built, and components have been used in the construction of the ZSU-23-4, ATS-59 and PTS. Although unsophisticated in design, uncomfortable, cramped and often badly finished, the combination of a powerful gun, excellent range and cross-country potential made this MBT series very advanced when first produced. All models have an infrared driving light and searchlight, and an NBC system. T-55s in service with the Indian Army have been upgraded either with the Soviet 115mm (4.52in) gun or British 105mm (4.13in) gun, whilst Israeli models have been substantially refitted and are armed with the 105mm (4.13in) L7-series gun.

T-34/85
MEDIUM TANK

Country of Origin: USSR
Crew: 5
Engine: V-2-34, V-12 diesel, water-cooled, developing 500hp at 1,800rpm or V-2-34 M V-12 diesel, water-cooled, developing 500hp at 1,800rpm
Performance: Speed 55km/h (34.16mls) (road); range 300km (186.34mls) (road); fuel 590 litres (129.96gal)
Armament: 1 × 85mm (3.37in) M1944 (ZIS-S53), elevation +25°, depression -5° (56 rounds carried); 2 × 7.62mm (0.3in) DT or DTM MGS-bow and A/A (2,394 rounds)

Armour: 18mm-75mm – (0.71in)-(2.97in)
Dimensions: Length 8.076m (26ft 5in) (inc gun), 6.19m (20ft 3in) (exc gun); width 2.997m (9ft 9in); height 2.743m (8ft 11in); ground clearance 0.38m (15in)
Weight: 32,000kg (70,400lb)

Ground Pressure: 0.83kg/cm^2 (11.87psi)
History: Considered by many experts to have been the leading tank design of the Second World War, the T-34/85 remained in production for many years. Although replaced in Soviet service in the late 1950s, the vehicle has been re-manufactured with new engines and wheels. This vehicle has been exported to many Third-World nations, and has seen action in the Middle East, Angola, Cyprus and Afghanistan.

CHALLENGER
MAIN BATTLE TANK

Country of Origin: United Kingdom
Crew: 4
Engine: Rolls-Royce CV12 TCA 12-cylinder 60V direct injection, 4-stroke diesel, compression-ignition

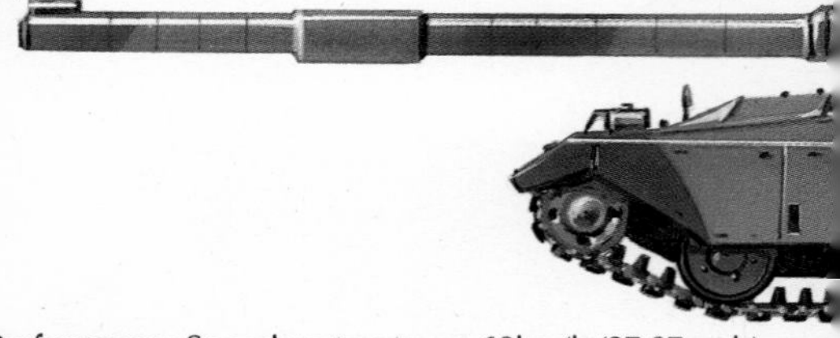

Performance: Speed up to approx 60km/h (37.27mph)
Armament: 120mm (4.75in) Tank gun L11
Dimensions: Length (gun forward) 11.55m (37ft 10in), length (gun rear) 9.86m (32ft 4in); height 2.88m (9ft 5in); width overall 3.52m (11ft 6in), over tracks 3.42m (11ft 2in); ground clearance 0.5m (1ft 7in)
Weight: Approx 62 tonnes (136,400lb) battle laden
Ground Pressure: 0.9kg/cm^2 (12.87psi)
History: The Challenger's main armament is the proven 120mm L11 semi-automatic gun. This fires bagged or combustible cased ammunition, loaded through a vertically sliding breech and initiated by an electrically primed vent tube. Storage allows up to 64 projectiles and 42 charge containers to be carried. Secondary armament is provided by two 7.62

machine guns, an L-8 coaxial machine gun and an L37 on the commander's cupola. The main gun is targeted with the aid of the Challenger's computerized fire-control system, which assimilates information on range, target movement, and present course and position before calculating main armament lay and bringing the gun to bear. Protection capability is enhanced by the use of British-designed Chobham armour, and a pressurized filtration system counters the NBC threat. A turret-mounted detector gives audio-visual warning if the tank is IR illuminated. The Challenger incorporates many new design concepts, and these, coupled with equipment such as the L11 gun, improved armour, and increased mobility, will take this important tank well beyond the year 2000.

CHIEFTAIN 900
MAIN BATTLE TANK
Country of Origin: UK

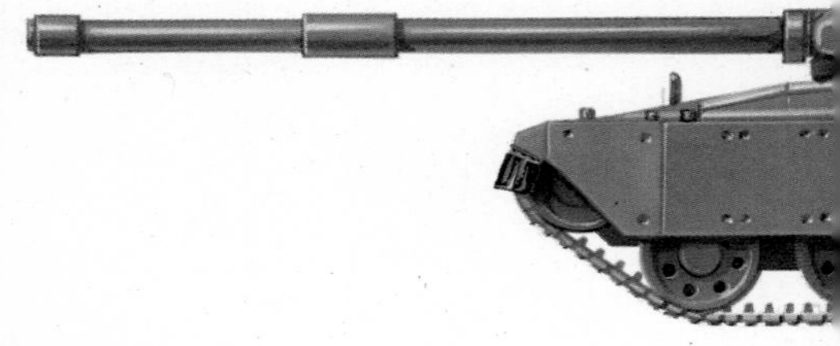

Crew: 4
Engine: Rolls-Royce 900E, 12 cylinder Condor, 60° direct injection turbo-charged 4-stroke diesel developing 900bhp at 2,300rpm
Performance: Speed 52km/h (32.3mph)
Armament: 1 × 120mm (4.72in) L11As gun; 1 × 7.62mm (0.3in) L8A2 MGs coaxial; 1 × 7.62mm (0.3in) L37A2 anti-aircraft and 2 × 5 smoke dischargers
Dimensions: Length: 7.52m (25ft 8in) hull; 10.8m (36ft 0in) gun forward; width 3.51m (11ft 8in); height 2.44m (8ft 1.5in) turret top

Weight: 56,000kg (123,200lbs)
Grand Pressure: 0.95kg/cm² (13.59psi)
History: The first prototype was shown at the British Equipment Exhibition in 1982 and although a second prototype was produced the Chieftain 900 has yet to enter production. Based on the Chieftain chasis the 900 features Chobham armour together with increased mobility and firepower. Main firepower is provided by the semi- automatic Royal Ordinance Factory 120mm (4.72in) rifled tank gun which is fitted with a fume-extractor.

CHIEFTAIN
MAIN BATTLE TANK

Country of Origin: United Kingdom
Crew: 4
Engine: Leyland L 60 No.4 Mark 7A turbo-charged

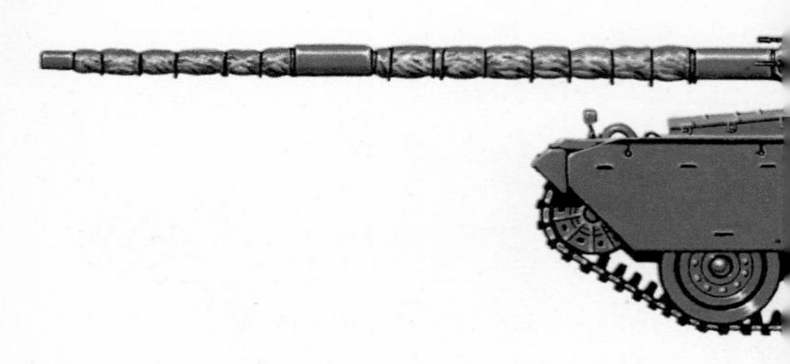

inline multi-fuel engine, developing 750 bhp at 2,250rpm
Performance: Speed 48km/h (29.81mph) (road); range 450km (279.5mls)
Armament: 1 × 120mm (4.75in) L11 A2 gun; 1 × 12.7mm (0.5in) ranging machine gun; 1 × 7.62mm (0.3in) coaxial machine gun; 1 × 7.62mm (0.3in) machine gun for A/A defence
Armour: 150mm (5.9in)
Weight: 55,000kg (121,000 lb) loaded
Dimensions: Length (gun forward) 10.795mm (35ft 4in); length (full) 7.52m (24ft 7in); width overall 3.66m (12ft); height 2.895m (9ft 5in); ground clearance 0.508m (1ft 7in)

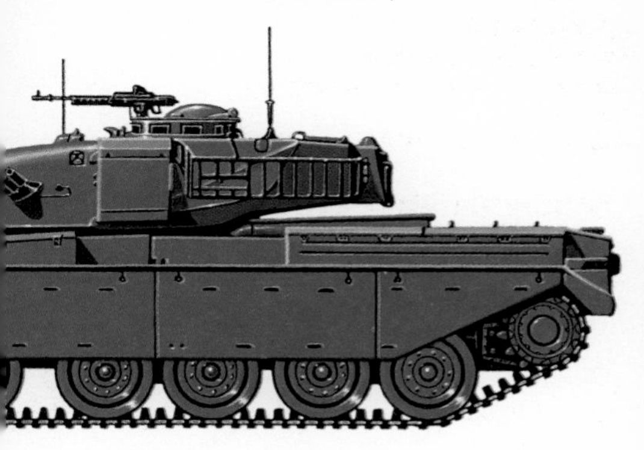

Ground Pressure: 0.90kg/cm² (12.87psi)
History: The Chieftain was first employed within BAOR in 1967, since which time nearly 1,000 Chieftains have been built for the British Army. Operated by a small joystick, and controllable either by the gunner or commander, the main armament can be brought to bear, whether or not the tank is on the move, within a few seconds. The turret can be traversed through 360° in approximately 20 seconds, elevation being +20° and depression –10°. The 120mm (4.72in) gun has a vertical breech block, fume extractor and thermal sleeve, and is capable of a sustained rate of fire of eight to ten rounds for the first minute and six rounds per minute thereafter.

VICKERS VALIANT MAIN BATTLE TANK

Country of Origin: United Kingdom
Crew: 4

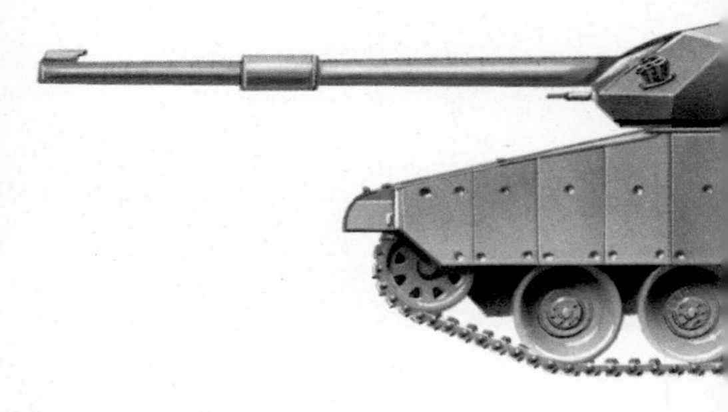

Engine: General Motors 12V-71T 12-cylinder diesel, developing 915bhp at 2,500rpm
Performance: Speed 59km/h (36.65mph) (road); range 603km (374.53mls) (road); fuel 1,000 litres (220.26gal)
Armament: 1 × 105mm (4.16in) gun, elevation +20°, depression -10° (60 rounds carried); 2 × 7.62mm (0.3in) MGs – coaxial and A/A (3,000 rounds carried); 2 × 6-barrelled smoke dischargers
Dimensions: Length 9.53m (31ft 3in) (gun forward), 7.51m (24ft 7in) (hull); width 3.3m (10ft 9in); height 2.64m (8ft 7in) (turret top), 3.24m (10ft 7in) (commander's sight); ground clearance 0.457m (18in)

Weight: 43,600kg (95,920lb) (loaded), 41,000kg (90,200lbs) (empty)
Ground pressure: 0.81kg/cm^2 (11.58psi)
History: Privately developed by Vickers Defence Systems Ltd for the export market, the Valiant first appeared at the British Army Equipment Exhibition in 1980. The hull is of welded aluminium, and a layer of Chobham armour can be added to the hull and turret in addition to side skirts to protect the tracks and suspension. The Valiant is fitted with a Cendor commander day/night sight, a laser rangefinder, and a Marconi SFC 600 fire-control system.

CENTURION MK10
MAIN BATTLE TANK

Country of Origin: United Kingdom
Crew: 4
Engine: Rolls-Royce Meteor Mk IV B 12-cylinder liquid-cooled petrol engine, developing 650bhp at 2550rpm
Performance: Speed 34.6km/h (21.5mph); range 190km (118mls)
Armament: 1 × 105mm (4.16in) Tank Gun; 1 × 7.62mm (0.3in) co-axial MG; 1 × 12.7mm (0.5in) RMG
Weight: 51,820kg (114,004lb) loaded
Dimensions: Length (gun forward) 9.854m (32ft 3in); length (full) 7.823m (25ft 7in); height 3.009m (9ft 10in); width overall 3.39m (11ft 1in); ground clearance 0.51m (1ft 8in)
Ground Pressure: 0.95kg/cm^2 (13.59psi)
History: First developed by AEC Ltd in 1944, the prototype tank just failed to see service in World War II. The Centurion first saw action in Korea and since

then has distinguished itself throughout the world, particularly in the Israeli Army. In its final form, the Mk 13, it was armed with the highly successful 105mm (4.13in) tank gun, also used in the Leopard 2, M60, Vickers MBT, Type 64 MBT and Swiss Pz 61 and Pz 68 tanks. The 105mm (4.13in) L7A2 gun is mounted in a turret capable of completing a 360° traverse in 26 seconds; elevation is +20° and depression –10°. Elevation and traverse are electric, with manual override in an emergency. In addition, the commander has the ability to override the gunner if necessary. The 105mm (4.13in) rifled tank gun is provided with a fume extractor on the barrel and has an effective range of 1,800m (1.14mls) when using APDS round or 3,000/4,000m (1.86/2.48mls) when using HESH. The gun is ranged using a 12.7mm (0.5in) ranging machine gun.

VICKERS
MAIN BATTLE TANK

Country of Origin: United Kingdom, although under production in India
Crew: 4

Engine: General Motors 12V 71T turbo-charged diesel, developing 800bhp at 2,500rpm
Performance: Mk3: Speed 53km/h (32.92mph) (road); range 600km (372.67mls)
Armament: 1 × 105mm (4.16in) tank gun; 1 × 12.7mm (0.5in) RMG; 1 × 7.62mm (0.3in) coaxial MG
Weight: 36,000kg (79,200lb) empty; 38,600kg (84,920lb) loaded
Dimensions (Mk3): Length (gun forward) 9.788m (32ft 1in); length (full) 7.561m (24ft 9in); height 2.71m (8ft 10in); width overall 3.168m (10ft 4in); ground clearance 0.406m (1ft 3in)
Ground Pressure: 0.87kg/cm^2 (12.44psi)

History: Originally designed in the late 1950s, the first prototype was sent to India in August 1963 and the first production model delivered in 1965. Since January 1969, approximately 500 Vickers MBTs have been produced in India and have entered service under the name 'Vijayanta'. A further 50 tanks were built by Vickers Ltd for export to Kuwait, entering service there between 1970 and 1972. A coaxial 7.62mm (0.3in) machine gun is mounted to the left of the main armament, there being provision for the mounting of a further machine gun on the commander's cupola if required. A six-barrelled smoke-discharger is mounted either side of the turret.

CROMWELL CRUISER TANK

Country of Origin: UK
Crew: 5
Engine: Rolls-Royce Meteor water-cooled vee-12, developing 600hp

Performance: Maximum speed 64km/h (40mph) (road); 29km/h (18mph) (cross country); Range 278km (137 miles) without external fuel tanks
Armament: (Mk IV) 1 x 74mm QF Mk V or Va gun; 1 x 7.92mm co-axial Besa machine-gun; 1 x 7.92mm Besa machine-gun in hull

Weight: 27,942kg (61,600lb)
Dimensions: Length 6.63m (21ft 9in); Width 2.9m (9ft 6in); Height 2.36m (7ft 9in)
Ground Pressure: 1kg/cm^2 (14.7psi)

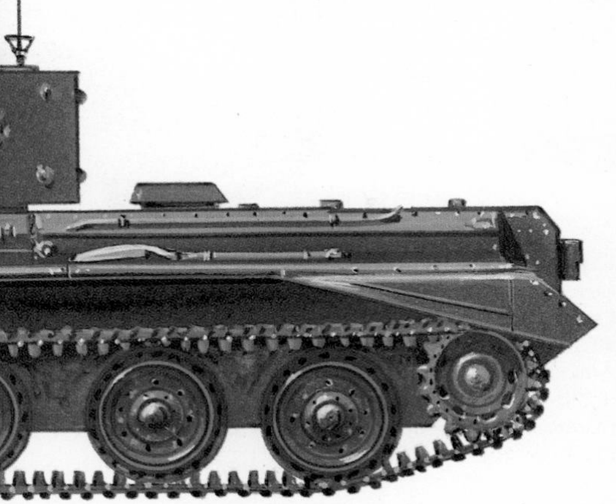

History: The first prototype of the Cromwell was produced in 1942 but the first production vehicles did not appear for another year. It was used in training during 1943/44 and first saw combat in the Normandy invasion, where it equipped the British 7th armoured Division. It was a reliable and popular tank and remained in British Army service until 1950.

M1 ABRAMS
MAIN BATTLE TANK

Country of Origin: USA
Crew: 4

Engine: Avco Lycoming AGT-1500 gas turbine, developing 1500hp at 3,000rpm
Performance: Speed 72.5km/h (45mph) (road), 48.5km/h (30mph) (cross country)
Armament: 1 × 105mm (4.16in) tank gun; 1 × 7.62mm (0.3in) coaxial machine gun; 1 × 12.7mm (0.5in) A/A machine gun (commander); 1 × 7.62mm (0.3in) A/A machine gun (loader)
Weight: 54,432kg (119,750lb) (loaded)
Dimensions: Length (gun forward) 9.766m (32ft); length (hull) 7.918m (25ft 11in); width overall 3.655m (11ft 11in); height 2.375m (7ft 9in) (turret roof), 2.895m (9ft 5in) (overall); ground clearance 0.482m (1ft 6in) (hull centre), 0.432m (1ft 5in) (hull sides)
History: It had long been accepted in the United

States that a new MBT was needed to counter the ever-improving Soviet arsenal. The M1 Abrams fulfills that need, and it will be the mainstay of US armoured units into the 21st Century. Ultimately the M1 Abrams will mount a German 120mm (4.72in) gun, but early models currently mount the established 105mm (4.13in) M-68 gun. The highly powered Avco Lycoming gas-turbine engine is coupled with the Detroit Diesel X-110-3B automatic transmission with four forward and two reverse gears. The engine has proved to be extremely noisy and hot and may well have to be adjusted in the future. Added protection is provided by an advanced armour construction similar to the British Chobham armour used on the Challenger and Leopard 2 MBTs.

TELEDYNE CONTINENTAL HIGH PERFORMANCE M60 MAIN BATTLE TANK

Country of Origin: USA
Crew: 4
Engine: A VCR-1790-1B 12-cylinder turbocharged die-

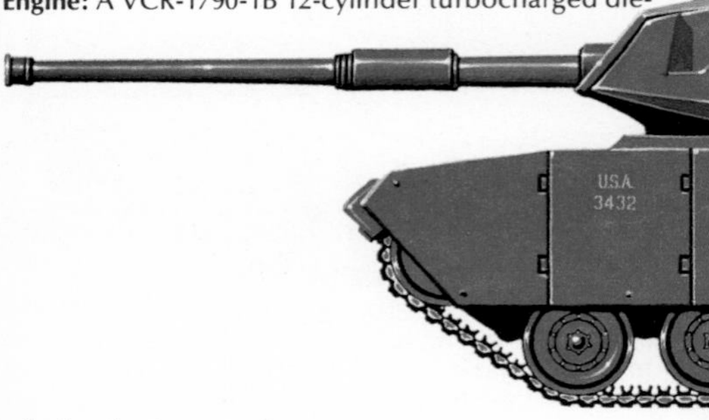

sel, developing 1,200hp at 2,400 rpm
Performance: Speed 74km/h (45.96mph) (road); range 500km (310.56mls) (road); fuel 1,457 litres (321gal)
Armament: 1 × 105mm (4.16in) M68 gun, elevation +20°, depression –9° (60 rounds carried); 1 × 7.62mm (0.3in) coaxial MG M73 (5,950 rounds); 1 × 12.7mm (0.5in) A/A MG M85 (900 rounds)
Armour: 25mm-110mm – (0.99in)-(4.36in)
Dimensions: Length 9.309m (30ft 6in) (gun forward), 6.946m (22ft 9in) (hull); width 3.631m (11ft 10in); height 3.213m (10ft 6in) (overall)

Weight: 46,266kg (101,785lb) (loaded), 42,184kg (92,805lb) (empty)
Ground Pressure: 0.78kg/cm^2 (11.15psi)
History: Developed as a private venture by the General Products Division of Teledyne, this is basically the M-60 series main battle tank fitted with a more powerful upgraded 12-cylinder turbo charged diesel engine, and Renk RK 340 fully automatic transmission. In addition, hydropneumatic suspension is fitted, and the vehicle is capable of accelerating from 0 to 32km/h (20mph) in 9 seconds.

M-60
MAIN BATTLE TANK

Country of Origin: USA
Crew: 4

Engine: Continental AVDS-1790-2A 12-cylinder air-cooled diesel developing 750bhp at 2,400rpm
Performance: 48.3km/h (30mph); range 500km (310mls)
Armament: 1 × 105mm (4.16in) tank gun; 1 × 7.62mm coaxial gun; 1 × 12.7mm (0.50) A/A machine gun
Weight: Loaded 46,266kg (101,785lb); unloaded 42,184kg (92805lb)
Dimensions: Length (gun forward) 9.309m (30ft 6in); length (hull) 6.946m (22ft 9in); width overall 3.631m (11ft 10in); height 3.213m (10ft 6in); ground clearance 0.463m (1ft 6in)
Ground Pressure: 0.783kg/cm² (11.20psi)
History: Developed from the M-47/M-48 family, the M-60 constitutes the main MBT in service with the United States at the moment. The M-60 entered service in 1959, being replaced in production by M-60A1 in 1962. The engine and Allison cross-drive transmis-

sion are at the rear of the hull, the latter with two forward and one reverse gears. Main armament consists of the British 105mm (4.13in) L7 gun, built under licence in the United States and designated the M68 gun. The turret is provided with an electro-hydraulic control system, with manual override, and is capable of a 360° traverse in fifteen seconds, with maximum gun elevation of +20° and depression of –10°. Average rate of fire is stated to be six to eight rounds per minute, with an effective range of 2,000m (1.26mls). A full NBC system, infrared lighting, a snorkel and fresh-air system are fitted as norm. The latest development, designated M-60A3, incorporates a number of improvements, such as full main armament stabilization. A Hughes laser rangefinder accurate to 5,000m (3.16mls), passive night-vision equipment, a solid-state M21 computer, and, thermal sleeve for the main armament.

M-48
MAIN BATTLE TANK

Country of Origin: USA
Crew: 4
Engine: AV-1790 5B/7/7B/7C Continental air-cooled engine, developing 750hp at 2,400rpm. Later models

fit AVDS-1790-2A/D Continental air-cooled engine, developing 750hp at 2,400rpm
Performance: Speed 48km/h (29.81mph); range 400km (248.45mls) (varies with model)
Armament: 1 × 90mm (3.56in) M41/L48 gun; 1 × 7.62mm (0.3in) coaxial machine gun; 1 × 12.7mm (0.5in) A/A machine gun
Weight: 44,906kg (98,793lb) (loaded), 42,240kg (92,928lb) (unloaded) in early models; 47,173kg (103,781lb) (loaded), 44,460kg (97,812lb) (unloaded)

in later models

Dimensions: Length (gun forward) 8.687m (28ft 5in); length (hull) 7.44m (24ft 4in); width overall 3.631m (11ft 10in); height 3.241m (10ft 7in) (early models), 3.086m (10ft 1in) (later models); ground clearance 0.393m (1ft 3in) (early models), 0.406m (1ft 3in) (later models)

Ground Pressure: 0.83kg/cm^2 (11.87psi)

History: The M-48 is essentially an improved M-47 with a revised turret and larger commander's cupola.

M-47
MEDIUM TANK

Country of Origin: USA
Crew: 5
Engine: Continental Model Av-1790-5B, 7 or 7B, V-12, 4-cycle, air-cooled petrol engine developing 810bhp at 2,800rpm

Performance: Speed 48km/h (29.81mph); range 128km (79.50mls)
Armament: 1 × 90mm (3.56in) M36 gun; 1 × 0.3 (7.62in) coaxial gun; 1 × 12.7mm (0.5in) A/A machine gun; 1 × 7.62mm (0.3in) machine gun in the bow
Weight: 44,707kg (98,355lb) (loaded), 42,130kg (92,686lb) (unloaded)
Dimensions: Length (gun forward) 8.553m (28ft); length (hull) 6.307m (20ft); width overall 3.51m (11ft 6in); height 3.016m (9ft 1in) (commander's cupola); ground clearance 0.469m (1ft 6in)

Ground Pressure: 0.935kg/cm^2 (13.37psi)
History: In many respects a modernized World War II Pershing, the M-47 was developed for, but did not see, service in the Korean War. Although soon replaced by the M-4b, some 8676 models of the M-47 were built, mostly for export under the United State Mutal Aid Programme. The main armament consists of the M36 90 mm (3.5in) rifled gun with either a T-shaped or cylindrical blast deflector. Mounted in a turret offering 360° travers in ten seconds, gun elevation of +19° and depression of-5° are possible.

M4A1 SHERMAN MEDIUM TANK

Country of Origin: USA

Crew: 5
Engine: R-975-C4, developing 400bhp at 2400rpm
Performance: Speed 39km/h (24mph) (road); range 160km (99.38mls) (cruising); fuel 651 litres (143.39gal)
Armament: 1 × 76mm (3in) gun, elevation +25°, depression -10° (71 rounds carried); 2 × 7.62mm (0.3in) MGs – coaxial and bow (6,250 rounds); 1 × 12.7mm (0.5in) A/A MG (600 rounds)
Armour: 12mm-75mm (0.48in - 2.97in)
Dimensions: Length 7.39m (24ft 2in); width 2.717m (8ft 10in); height 3.425m (11ft 2in); ground clearance 0.43m (17in)

Weight: 32,044kg (70,497lb) (loaded)
Ground Pressure: 1.02kg/cm^2 (14.59psi)
History: The most famous Allied tank of World War II, the M4 Sherman continues in service, particularly with the US and Canadian armies. Replaced quickly by major armies in the postwar period, the Sherman was exported to many countries. In Israel many modifications have been made to the original design. Seeing action in all the Arab-Israeli wars, many of the Sherman 'gun-tanks' have now been further modified into self-propelled mortars, armoured ambulances, and engineer vehicles.

JAGDPANZER SK 105
LIGHT TANK/TANK DESTROYER

Country of Origin: Austria
Crew: 3
Engine: Steyr Model 7FA 6-cylinder diesel, developing

320hp at 2,300rpm
Performance: Speed 65km/h (40mph) (road); range 520km (323mls); fuel 400 litres (88.11gal)
Armament: 1 × 105mm (4.16in) gun, elevation +13°, depression -8° (44 rounds carried); 1 × 7.62mm (0.3in) coaxial MG (2,000 rounds carried); 2 × 3 smoke dischargers
Armour: 8mm-40mm (0.32in - 1.58in)
Dimensions: Length 7.763m (25ft 5in) (inc gun), 5.58m (18.3in) (exc gun); width 2.05m (6ft 8in); height 2.514m (8ft 2in); ground clearance 0.40m (15.75in)

Weight: 17,500kg (38,500lb) (loaded)
Ground Pressure: 0.68kg/cm² (9.72psi)
History: The first prototype developed by Saurer was completed in 1967, with a second in 1969. The pre-production models appeared in 1971 after Saurer had been taken over by Steyr-Daimler-Püch, and production has continued steadily since. The 105mm (4.16in) gun fires the same ammunition as the French AMX-30, including HEAT. The laser rangefinder is fitted at the rear of the turret roof beneath an infrared/white light searchlight.

X1A1
LIGHT TANK

Country of Origin: Brazil
Crew: 4
Engine: Scania 6-cylinder diesel, developing 280hp

Performance: Speed 60km/h (37.27mph); range 520km (323mls)
Armament: 1 × 90mm (3.56in) gun; 1 × 7.62mm (0.3in) coaxial MG; 1 × 12.7mm (0.5in) MG; 2 × 3 smoke dischargers
Dimensions: Length 6.36m (20ft 10in) (gun forward), 5.3m (17ft 4in) (hull); width 2.4m (7ft 10in); height 2.45m (8ft) (turret top); ground clearance 0.5m (20in)
Weight: 17,000kg (37,400lb)
Ground Pressure: 0.55kg/cm² (7.87psi)

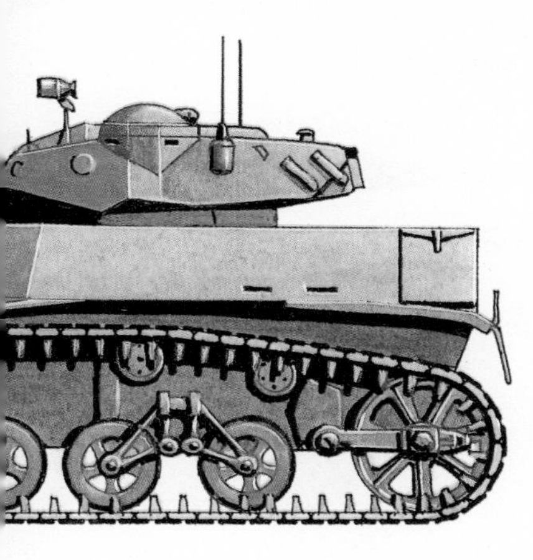

History: The Bernardini Company rebuilt 80 American-supplied M3A1 Stuart light tanks designated X1A, the extensive rebuild including new armour above the tracks, new turret with a French D-921 F1 90mm (3.5in) gun, a new fire-control system, and a Scania diesel engine. Weighing in at 15,000kg (33,000lb), the X1A had a range of 450km (280mls). Further development resulted in the X1A1, which was not ordered by the Brazilian Army but is available for export.

X1A2
LIGHT TANK

Country of Origin: Brazil
Crew: 3

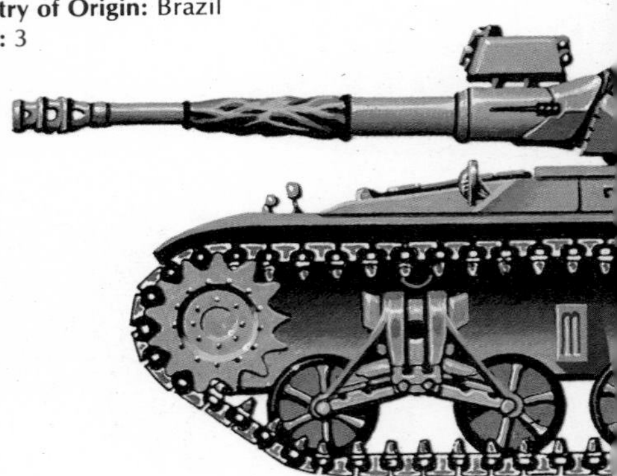

Engine: Scania model 0S-11 6-cylinder water-cooled turbo-charged diesel, developing 300hp at 2,200rpm
Performance: Speed 55km/h (33.16mph) (road); range 600km (372.67mls); fuel 600 litres (132gal)
Armament: 1 × 90mm (3.56in) gun, elevation +17°, depression -8° (66 rounds carried); 1 × 7.62mm (0.3in) coaxial MG (2,500 rounds), 1 × 12.7mm (0.5in) A/A MG (750 rounds); 3 smoke dischargers either side of turret
Dimensions: Length 7.1m (23ft 3in) (gun forward),

6.5m (21ft 3in) (hull); width 2.6m (8ft 6in); height 2.45m (8ft) (turret top); ground clearance 0.5m (20in)
Weight: 19,000kg (41,800lb) (loaded)
Ground Pressure: 0.63kg/cm² (9.01psi)
History: Now in production for the Brazilian Army, the X1A2 is a completely new vehicle. Prototypes were armed with a French 90mm (3.5in) D-921 gun, but production models switched to the Belgian Cockerill, which is now manufactured under licence in Brazil by ENGESA.

TYPE 62
LIGHT TANK
Country of Origin: China

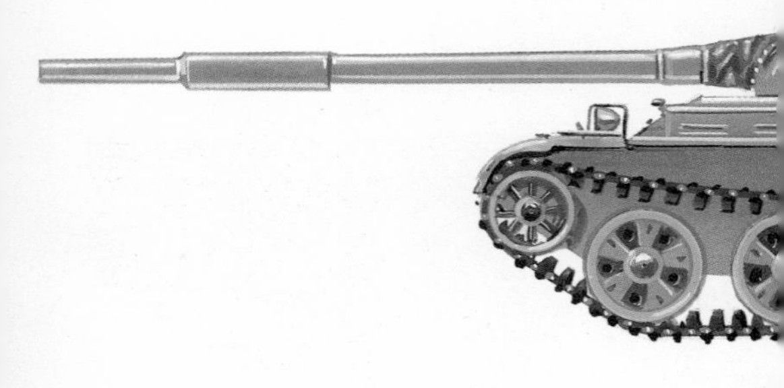

Crew: 4
Engine: V-12 diesel, developing 520hp at 2,000rpm
Performance: Speed 50km/h (31.06mph) (road), 9km/h (5.59mph) (water); range 300km (186.34mls); fuel 545 litres (120gal)
Armament: 1 × 85mm (3.37in) gun, elevation +18°, depression -5°; 1 × 7.62mm (0.3in) coaxial MG; 1 × 12.7mm (0.5in) A/A MG
Armour: 14mm (0.55in)
Dimensions: Length 8.27m (27ft 1in) (gun forward), 6.91m (22ft 7in) (hull); width 3.25m (10ft 7in); height

2.19m (7ft 2in) (turret roof); ground clearance 0.37m (14.5in)
Weight: 18,000kg (39,600lb) (loaded)
History: This vehicle is a Chinese development of the Soviet PT-76 reconnaissance tank, of which many were supplied to the People's Liberation Army in the 1950s and 60s. The original Soviet vehicle has been re-engined and up-gunned with a 85mm (3.35in) gun. Currently the standard Chinese light tank, the Type 62 has been exported to many Southeast Asian countries, and to some African states.

AMX-13 LIGHT TANK

Country of Origin: France
Crew: 3
Engine: SOFAM Model 8 GX6 8-cylinder, water-cooled, petrol engine developing 250hp at 3,200rpm
Performance: 60km/h (37.27mph); range 350/400km (217/248mls)

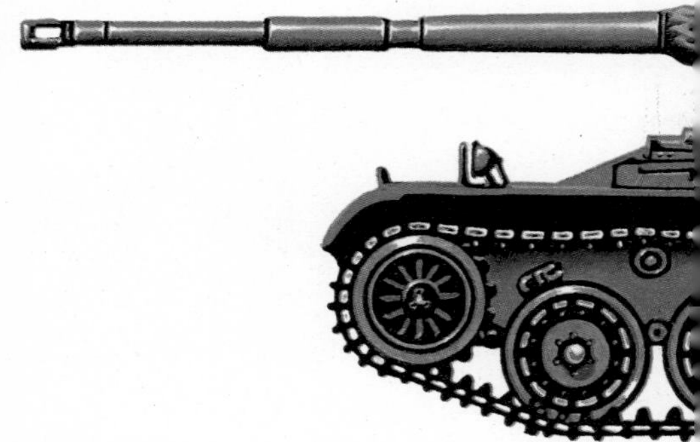

Armament: 1 × 90mm (3.56in) gun; 1 × 7.62mm (0.3in) coaxial gun
Weight: 13,000kg (28,600lb) (empty), 15,000kg (33,000lb) (loaded)
Dimensions: Length (gun forward) 6.38m (20ft 11in); length (hull) 4.88m (16ft); width overall 2.5m (8ft 2in); height 2.3m (7ft 6in); ground clearance 0.37m (1ft 2in)

Ground Pressure: 0.76kg/cm² (10.87psi)
History: The AMX-13 was adopted into French service in 1953, since which time over 4,000 have been built. In addition, several thousand adaptations of the basic vehicle have been built, including 105mm (4.13in) and 155mm (6.10in) self-propelled guns, armoured personnel carriers, and anti-aircraft vehicles.

PT-76
AMPHIBIOUS RECONNAISSANCE VEHICLE
Country of Origin: USSR

Crew: 3
Engine: Model V6, 6-cylinder in-line water-cooled diesel engine developing 240hp at 1,800rpm
Performance: Speed 44km/h (27.33mph) (road), 11km/h (6.83mph) (water); range 255km (158.39mls)
Armament: 1 × 76.2mm (3in) D-56 gun; 1 × 7.62mm (0.3in) SGMT coaxial machine gun
Weight: 14,000kg (30,800lbs) loaded
Dimensions: Length (gun forward) 7.625m (25ft); length (hull) 6.91m (22ft 7in); width overall 3.16m (10ft 4in); height 2.2m (7ft 2in); ground clearance 0.37m (1ft 2in)
Ground Pressure: 0.48kg/cm^2 (6.86psi)
History: Designed initially as a primary reconnais-

sance vehicle, the PT-76 entered service in the Soviet Union in 1952 and was in service throughout the

Warsaw Pact by 1976. The PT-76 has no NBC or night-vision equipment. It has seen service in both the Middle and Far East but is vulnerable against more modern armament and has now been replaced as a front-line reconnaissance vehicle by the BMP. It has been retained for use in the Soviet Union within the Naval Infantry, where its amphibious qualities can best be exploited. The engine is one half of that fitted to the T-54 MBT; it has a gear box with four forward and one reverse gear, and is fitted with a pre-heater to facilitate cold weather starting. Propulsion in water is via two water jets at the rear of the hull, a trim vane being erected on the front of the hull before entering the water.

M41
LIGHT TANK

Country of Origin: USA
Crew: 4

Engine: M41 and M41A1, Continental or Lycoming AOS-895-3; M41A2 and M41A3, Continental or Lycoming 1051-895-5, 6-cylinder, air-cooled petrol engine, super-charged, developing 500hp at 2,800rpm
Performance: Speed 72km/h (44.72mph) (road); range 161km (100mls); fuel 530 litres (116.74gal)
Armament: 1 × 76mm (3in) M32 gun, elevation +19°, depression -10° (57 rounds carried); 1 × 7.62mm (0.3in) M1919A4E1 coaxial MG (5,000 rounds); 1 × 12.7mm (0.5in) M2 A (2,175 rounds)
Armour: 12mm-38mm (0.48in - 1.50in)
Dimensions: Length 8.212m (26ft 11in) (inc gun), 5.819m (19ft 1in) (exc gun); width 3.198m (10ft 5in);

height 3.075m (10ft 1in) (with MG), 2.726m (8ft 11in) (w/o MG); ground clearance 0.45m (17.5in)
Weight: 23,495kg (51,689lb) (loaded)
Ground Pressure: 0.72kg/cm² (10.30psi)
History: The M41 light tank was for many years the standard reconnaissance vehicle of United States armoured regiments, where it replaced the earlier M24 Chaffee. The vehicle, since replaced in US service, has been vigorously exported, and today is an important vehicle for many nations, particularly those where the larger MBTs are unsuitable. The tank remains in service in Brazil, Thailand, Taiwan, Denmark and Spain, who have all continually updated the vehicle.

M551 GENERAL SHERIDAN
LIGHT TANK/RECONNAISSANCE VEHICLE

Country of Origin: USA

Crew: 4

Engine: Detroit Diesel 6V53T, developing 300hp at 2,800rpm

Performance: Speed 70km/h (43.48mph) (road), 5.8km/h (3.6mph) (water); range 600km (372.67mls) (road); fuel 598 litres (131.72gal)
Armament: 1 × 152mm (6.02in) M81 launcher, elevation +19.5°, depression -8° (20 conventional rounds, 10 Shillelagh missiles carried); 1 × 7.62mm (0.3in) M73 coaxial MG (3,080 rounds); 1 × 12.7mm (0.5in) M2 MG at commander's cupola (1,000 rounds); 8 grenade launchers (four either side of turret)
Dimensions: Length 6.299m (20ft 7in); width 2.819m (9ft 2in); height 2.946m (9ft 7in) (inc MG); ground clearance 0.48m (19in)
Weight: 15,830kg (34,826lb) (loaded), 13,589kg (29,896lb) (empty)

Ground Pressure: 0.49kg/cm^2 (7.01psi)
History: The first prototype was completed in 1962 and designated XM551. First production vehicle from the Allison Division of General Motors was completed in 1966, and 1,700 vehicles had been manufactured when production ceased in 1970. The M551 replaced the M41 with the US Army, but the M551 itself only remains in service with the 82nd Airborne Division. The Sheridan's weapon system comprises either a launcher for Shillelagh missiles with a range of 3,000m (1.86mls), or conventional rounds including HEAT-T-MP, white phosphorus, TP-T and canister.

M24 CHAFFEE LIGHT TANK

Country of Origin: USA
Crew: 4/5
Engine: 2 × Cadillac Model 44T24 petrol. V-8, water-cooled, developing 110hp at 3,400rpm (each)
Performance: Speed 55km/h (34.16mph) (road); range

281km (174.53mls) (road) 173km (107.45mls) (cross country); fuel 416 litres (91.63gal)
Armament: 1 × 75mm (2.97in) M6 gun, elevation +15°, depression -10° (48 rounds carried); 1 × 12.7mm (0.5in) M2 A/A MG (440 rounds); 2 × 7.62mm (0.3in) MGS – coaxial and bow (3,750 rounds)
Armour: 10mm-38mm – (0.4in)-(1.5in)
Dimensions: Length 5.486m (17ft 11in) (inc gun), 5.028m (16ft 5in) (exc gun); width 2.95m (9ft 8in); height 2.77m (9ft 1in) (inc MG), 2.46m (8ft) (commander's cupola); ground clearance 0.457m (18in)
Weight: 18,370kg (40,414lb) (loaded), 16,440kg

(36,168lb) (empty)
Ground Pressure: 0.78kg/cm² (11.15psi)
History: Probably the most effective of the light tank designs, the M24 Chaffee saw action in the final stages of the Second World War and remains, albeit in modernized form, in front-line service today. Noted for its speed and manoeuvrability, the M24 was the vehicle that managed to hold up the North Korean invasion of the South in 1950, where it was battling against the T-34/85. Although replaced in US service, substantial numbers remain in service with the Greek, Turkish, Taiwanese and other armies.

M3 STUART
LIGHT TANK

Country of origin: USA
Crew: 4
Engine: Continental W670-9A, 7-cylinder petrol engine, developing 250hp at 2,400rpm
Performance: Speed 56km/h (34.78mph) (road); range 120km (74.53mls) (cruising); fuel 212 litres (46.7gal)
Armament: 1 x 37mm (1.47in) gun, elevation +20°, depression -10° (108 rounds carried); 3 x 7.62mm (0.3in) MGS - coaxial, bow and A/A (6,890 rounds)

Armour: 10mm-44mm (0.4in - 1.74in)
Dimensions: Length 4.54m (14ft 10in); width 2.24m (7ft 4in); height 2.30m (7ft 6in); ground clearance 0.42m (16.5in)
Weight: 12,927kg (28,439lb) (loaded)

Ground Pressure: 0.74kg/cm² (10.58psi)
History: The M3 saw service with the British Army from 1941 and was a direct development from the M2. A later development was the M3A3, which had an all-welded hull, a fuel capacity of 386lit (85gal), sand shields and additional ammunition storage made possible by lengthened sponsors at the rear of the vehicle. Modified M3s remain in service in South America only.

M5A1
LIGHT TANK

Country of Origin: USA
Crew: 4
Engine: Two V-8 Cadillac Series 42 petrol engines, developing 110hp at 3,200rpm
Performance: Speed 58km/h (36mph) (road); range 160km (99.38mls) (cruising); fuel 310 litres (68.28gal)
Armament: 1 × 37mm (1.47in) gun, elevation +20°, depression -10° (147 rounds carried); 3 × 7.62mm (0.3in) MGs – coaxial, bow and A/A (6,500 rounds)
Armour: 10mm-63mm – (0.4in - 2.49in)
Dimensions: Length 4.84m (14ft 8in); width 2.29m (7ft 6in); height 2.30m (7ft 6in); ground clearance 0.35m (14in)
Weight: 15,397kg (33,873lb) (loaded)
Ground Pressure: 0.88kg/cm^2 (12.58psi)

History: Originally designated M3E2, this development of the M3 was redesignated M5A1 to prevent confusion. It was powered by two V-8 Cadillac Series 42 petrol engines. An improved turret was incorporated into late M5A1 models, these having a radio bulge at the rear and an A/A gun on the right hand side. Another variant, the M8, was fitted with a new turret mounting a 75mm (3in) howitzer with an elevation of +40° and a depression of –20°. In French service the vehicle was used extensively in that nation's colonial wars, but is now only found in South American service.

ENGESA EE-9 CASCAVEL ARMOURED CAR

Country of Origin: Brazil
Crew: 3

Engine: Mercedes-Benz OM-352A 6-cylinder water-cooled turbocharged diesel, developing 190hp at 2,800rpm
Performance: Speed 100km/h (62.11mph); range 1,000km (621.12mls); fuel 360 litres (79.3 gal)
Armament: 1 × 90mm (3.56in) gun, elevation +15°, depression -8° (45 rounds carried); 2 × 7.62mm (0.3in) MGs – one coaxial and A/A (optional) – (2,400 rounds); 2 × 2 smoke dischargers
Armour: 12mm (0.48in) (max)
Dimensions: Length 6.22m (20ft 4in) (gun forward), 5.19m (17ft) (hull); width 2.59m (8ft 5in); height 2.36m (7ft 8in) (commander's cupola); ground clearance 0.375m (14.75in) front axle

Weight: 12,200kg (26,840lb) (loaded), 11,800kg (25,960lb) (empty)
History: Production vehicles were available in 1974 following four years of prototypes and pre-production models. Developed for the Brazilian Army, the EE-9 has many automative components in common with a EE-11 Urutu APC. Variants include the Mark 1 with 37mm (1.45in) gun; Mark II, an export model no longer in production, which was fitted with a French H-90 turret with a 90mm (3.54 in) gun and automatic transmission; Mark III, fitted with an ENGESA ET-90 turret; and Mark IV, similar to the Mark III but with a General Motors Diesel 6V53 engine.

ENGESA EE-3 JARARACA SCOUT CAR

Country of Origin: Brazil
Crew: 3

Engine: Daimler-Benz OM-314 4-cylinder water-cooled diesel, developing 120hp at 2,800rpm
Performance: Speed 90km/h (56mph); range 750km (466mls); fuel 135 lit (30gal)
Armament: 1 × 12.7mm (0.5in) MG
Armour: Hull consists of an outer layer of hard steel and an inner layer of softer steel roll-bonded and heat treated for maximum protection
Dimensions: Length 4.195m (13ft 9in); width 2.13m (6ft

11in); height 1.56m (5ft 1in) (w/o armament); ground clearance 0.315m (12.5in)
Weight: 5,200kg (11,440lb) maximum

History: Designed to meet the requirements of the Brazilian Army, the ENGESA EE-3 Jararaca entered production in 1980. It can be fitted with a wide range of armament including a 7.62mm (0.3in) or 12.7mm (0.5in) MG, a 20mm (0.8in) cannon, a 60mm (2.4in) breech-loaded mortar, a 106mm (4.16in) M40 recoilless gun, or a MILAN ATGW. In addition to the armament, optional equipment includes passive night vision and an intercom system.

AMX-10RC
RECONNAISSANCE VEHICLE

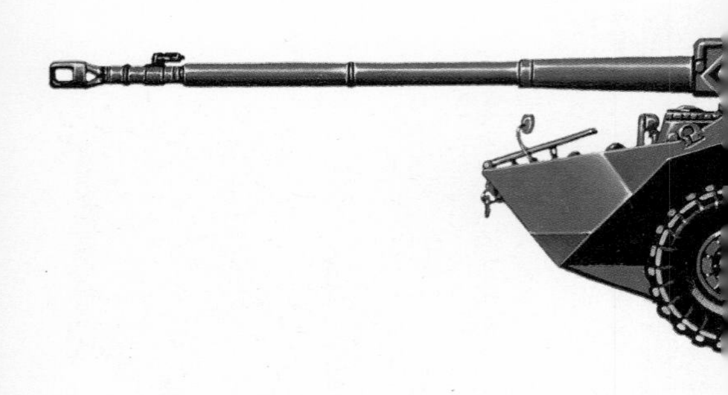

Country of Origin: France
Crew: 4
Engine: Hispano-Suiza HS-115 8-cylinder supercharged diesel, developing 280hp at 3,000rpm
Performance: Speed 85km/h (52.8mph) (road), 7km/h (4.35mph) (water); range 800km (496.89mls)
Armament: 1 × 105mm (4.16in) gun; 1 × 7.62mm (0.3in) CA; 2 × 2 smoke dischargers
Dimensions: Length 6.35m (20ft 9in); width 2.86m (9ft 4in); height 2.68m (8ft 9in)
History: The AMX-10RC is a reconnaissance vehicle designed for antitank combat. It is amphibious with-

out preparation, and has very good mobility. Its six wide wheels, and oil and air suspension, coupled with variable ground clearance, afford the cross-country capability usually associated with tracked vehicles. The 105mm (4.13in) gun fires a high velocity, hollow-charged shell, and is targeted by a high-performance fire-control system, with a ×10 magnification laser rangefinder and automatic fire-correction control, fitted with a low-light TV for night operation. This vehicle is derived from the AMX-10P, and is in service with the French Army.

EBR 75
HEAVY ARMOURED CAR

Country of Origin: France
Crew: 4

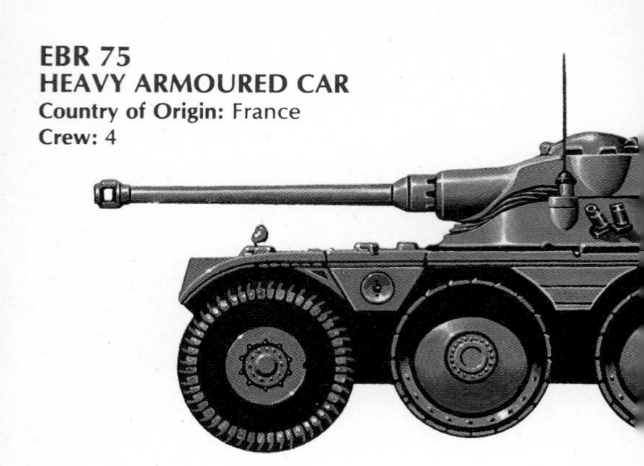

Engine: Panhard 12-cylinder petrol engine, developing 200hp at 3,700rpm
Performance: Speed 105km/h (65.22mph) (road); range 650km (403.73mls); fuel 380 litres (83.7gal)
Armament: 1 × 75mm (2.97in) gun, elevation +15°, depression -10° (56 rounds carried); 1 × 7.5mm (0.3in) MG coaxial
Armour: 10mm-40mm (0.4in)-(1.58in)
Dimensions: Length 6.15m (20ft 2in) (o/a FL-11 turret); 5.56m (18ft 2in) (vehicle only); width 2.42m (7ft 11in); height 2.32m (7ft 7in) (FL-11 on 8 wheels); 2.24m (7ft 4in) (FL-11 on 4 wheels); ground clearance 0.41m (16.14in) on 8 wheels, 0.33m (13in) on 4 wheels

Weight: 13,500kg (29,700lb) (loaded FL-11)
Ground Pressure: 0.75kg/cm² (10.73psi) (on 8 wheels)
History: Designated Panhard Model No. 201 in 1939, the plans for this vehicle were destroyed during

World War II. Development work was not resumed until after the War, with a prototype being ready in 1948. Production commenced in 1950 and ceased 10 years later, after 1,200 had been built. The four-man crew consists of commander, gunner and two drivers. The EBR 75 is fitted with either a FL-11 or FL-10 turret, although the variants with FL-10 turrets have now been withdrawn and those with FL-11 turrets are slowly being replaced by the AMX-10RL (6 × 6). Final production models were fitted with a 90mm (3.56in) gun in an FL-11 turret, firing fin-stabilized ammunition. Being replaced in French service, the EBR is still used by Portugal and Morocco.

RENAULT VBC 90
ARMOURED CAR
Country of Origin: France

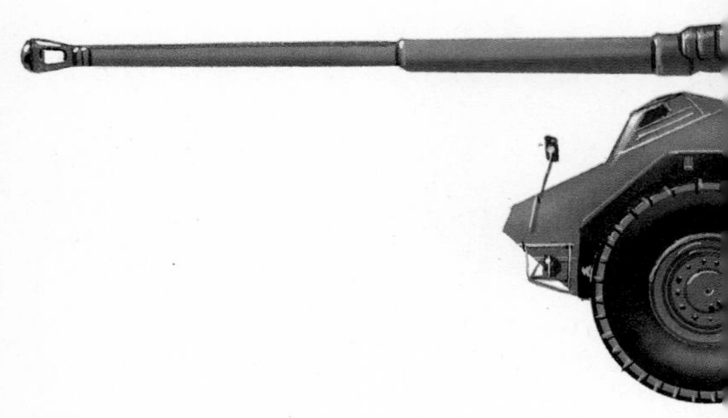

Crew: 3
Engine: MAN D.2356 HM 72 in-line air-cooled diesel, developing 235hp at 2,200rpm
Performance: Speed 112.7lkm/h (70mph); range 1127km (700mls)
Armament: 1 × 90mm (3.56in) gun; 1 × 7.62mm (0.3in) CA; 1 × 7.62mm AP
Weight: 12.8tonnes (28,160lbs)
Dimensions: Length 5.495m (18ft); width 2.49m (8ft 2in); height 2.55m (8ft 4in)

History: This highly mobile six-wheeled, antitank, armoured fighting vehicle is derived from the VAB series. Its L GIAT 7S 90 turret houses an accurate, high-performance 90mm (3.54in) gun firing APFSDS, hollow-charge, and HE ammunition, with a range of 1,700m (1.05mls). The VBC 90 carries a total of 45 rounds and the gun has an effective time-control system, linked with a SOPTAC computer. The VBC 90 has been adopted by several countries including France.

HOTCHKISS SP. 1A
RECONNAISSANCE VEHICLE
Country of Origin: France

Crew: 5
Engine: Hotchkiss 6-cylinder, OHV, water-cooled petrol engine developing 164hp at 3,900rpm
Performance: Speed 58km/h (36mph) (road); range 390km (242.24mls); fuel 330 litres (72.69 gal)
Armament: 1 × 20mm (0.8in) cannon, elevation +75°, depression -10° (500 rounds carried)
Armour: 8mm-15mm (0.32in - 0.59in)
Dimensions: Length 4.51m (14ft 9in); width 2.28m (7ft 5in); height 1.97m (6ft 5in); ground clearance 0.35m (13.77in)

Weight: 8,200kg (18,040lb) (loaded)
Ground Pressure: 0.58kg/cm² (8.29lb/sq in)
History: Developed in France, but never used by the French armed forces, the Hotchkiss series of carriers became an important constituent of the re-formed West German Army in the mid-1950s. Supplied in many versions, including reconnaissance, observation, engineer and ambulance versions, the vehicle has now been replaced by more recent construction. However, substantial numbers are probably held in reserve.

PANHARD AML
RECONNAISSANCE VEHICLE (WHEELED 4 × 4)

Country of Origin: France
Crew: 3
Engine: Panhard Model 4 HD, 4-cylinder air-cooled petrol engine developing 90hp at 4,700 rpm
Performance: Speed 90km/h (55.9mph) (road); range 600km (372.67mls)
Armament: 1 × 90mm (3.56in) and 1 × 7.62mm (0.3in) CA. 2 smoke dischargers either side of turret
Armour: 8mm-12mm (0.32in-0.47in)
Dimensions: Length 3.79m (12ft 5in); width 1.97m (6ft

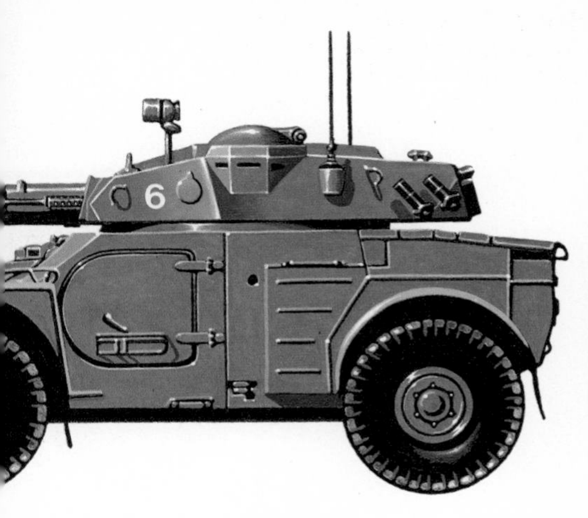

5in); height 2.07m (6ft 9in)
Weight: 5.5 tonnes (12,100lb)
History: The AML 60/20 is armed with a 60mm HB 60LP mortar gun with 20mm (0.79in) M693 auto cannon housed in a Hispano-Suiza 'Serval' turret. The AML 90 is armed with a 90mm (3.54in) F1 gun, with one 7.62mm (0.3in) MG mounted coaxially and a second mounted forward of the cupola for close defence. The AML 60/20 and the AML 90 share the same chassis, engine, and electrical and mechanical systems.

APE
AMPHIBIOUS ENGINEER RECONNAISSANCE VEHICLE

Country of Origin: West Germany
Crew: 5

Engine: Mercedes-Benz OM 403 V-8 water-cooled diesel developing 390hp at 2,500rpm
Performance: Speed 83km/h (51.55mph) (road), 12km/h (7.45mph) (water); range 800km (496.89mls) (cross-country)
Armament: 1 × 20mm (0.8in) cannon; 6 smoke dischargers
Dimensions: Length 6.93m (22ft 8in); width 3.08m (10ft 1in); height 1.4m (7ft 10in) (hull top); ground clearance 0.485m (19.1in) hull

Weight: 14,500kg (31,900lb) (loaded)
History: Only available as a prototype, production of the APE is currently pending due to budgetry prob-

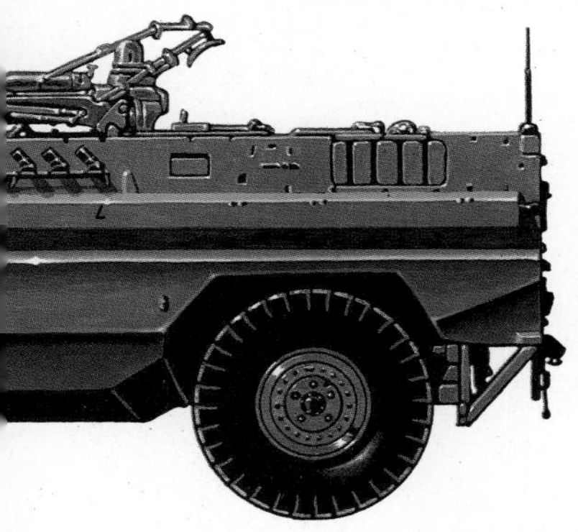

lems. When afloat, the axles of the APE are raised hydraulically to reduce drag, and power is provided by two rear-mounted propellers that can be turned through 360°, thereby affording excellent manoeuvrability. The APE has balloon tyres, the pressure of which can be varied by the driver to suit ground conditions. The first prototype was completed in 1977 using almost all of the same automatic parts as the Transportpanzer I multipurpose 6 × 6 vehicle now in service with the West German Army.

LUCHS (LYNX) SPAHPANZER 2
WHEELED ARMOURED RECONNAISSANCE VEHICLE

Country of Origin: West Germany
Crew: 4
Engine: Daimler-Benz 10-cylinder multi-fuel engine, 390hp water-cooled
Performance: Speed 90kph (55.90mph); range 800km (496.89mls)
Armament: 20mm (0.8in) cannon; 7.62mm (0.33in) machine gun

Weight: 19.5 tonnes (42,900lb)
Dimensions: Length 7.743m (25ft 4in); height 2.90m (9ft 6in); width 2.98m 9ft 9in)

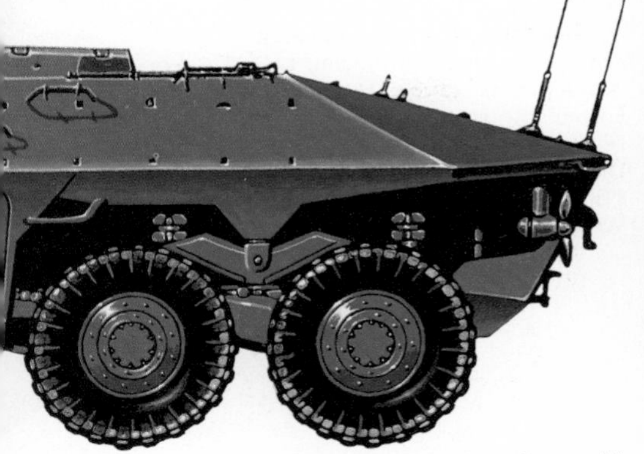

History: The Luchs was produced by Rheinstahl between 1975 and 1978, to a Mercedes-Benz design, as a replacement for the Bundeswehr's ageing M41s and Hotchkiss SPz 11-2 reconnaissance vehicles. The vehicle is highly mobile, having eight-wheel steering when required off-road (four-wheel steering being employed normally). It is also fully amphibious, two propellers giving a water speed of 9kp/h (5.6mph).

WIESEL
AIR-PORTABLE VEHICLE

Country of Origin: West Germany
Crew: 3
Engine: Audi water-cooled petrol, developing 100hp
Performance: Speed 85km/h (52.8mph); range 200km (124.22mls); fuel 80 litres (17.62gal)
Armament: 1 × 7.62m (0.3in) MG; 1 × Rheinmetall 20mm (0.8in) cannon; Hughes TOW ATGW system; Euromissile HAKO turret with two HOT ATGW in ready-to-launch position
Dimensions: Length 3.26m (10ft 8in); width 1.82m (5ft 11in); height 1.84m (6ft)
Weight: 2,600kg (5,720lb)

Ground Pressure: 0.35kg/cm² (5.01psi)
History: This specialist air-portable vehicle was designed at the request of West Germany's airborne units. The vehicle, bearing a resemblance to the WWII British Bren-Gun Carrier, is designed to take a choice of weapons fits, of which the antitank TOW

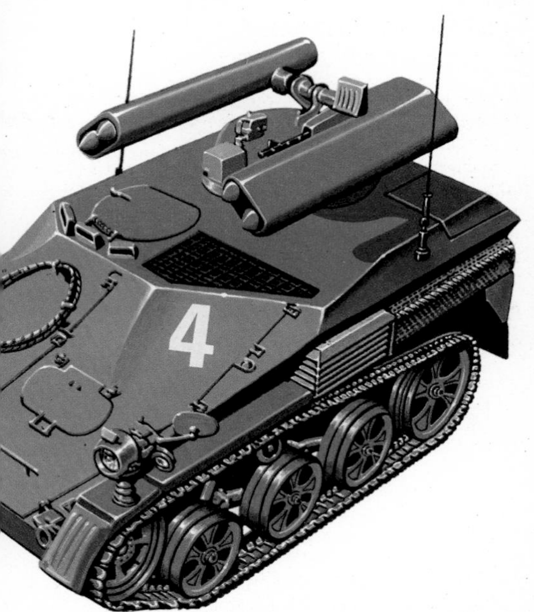

missile system will be the most common. Now in production, the Wiesel is also being offered for export.

RAM V-1
LIGHT ARMOURED RECONNAISSANCE VEHICLE

Country of Origin: Israel
Crew: 2 + 4
Engine: Deutz F6L-912 air-cooled diesel developing 115hp
Performance: Speed 95km/h (59mph); range 850km (528mls) (road); fuel 120 litres (26.43gal)
Armament: Three 7.62mm (0.3in) MGs, six rifles, one 52mm (2.05in) mortar plus grenades, flares and ammunitions

Dimensions: Length 5.02m (16ft 5in); width 21.03m (6ft 7in); height 1.59m (5ft 2in) (loaded), 1.71m (5ft 7in) (unloaded)
Weight: 4,100kg (9,020lb) (unloaded)

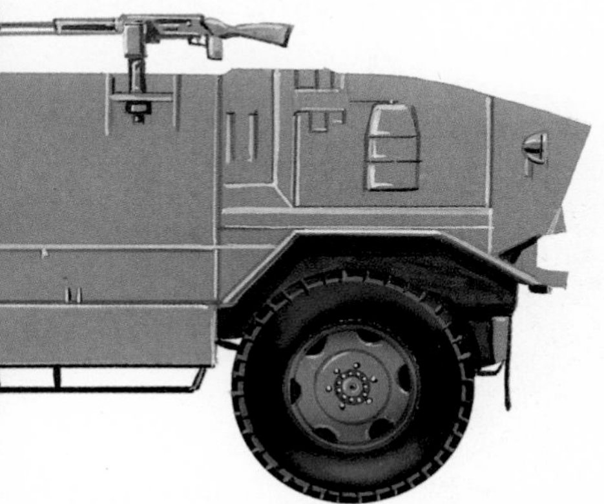

History: Designed for the Israeli Army as a lightweight reconnaissance vehicle and based on the combat experience of many campaigns; this vehicle is now in service with not only the Israeli forces but also with some South American and African States. The vehicle in its scout role is fitted with standard MAG (FN) machine guns, but it can also be outfitted as a tank destroyer equipped with the Hughes TOW missile system.

RBY Mk1
ARMOURED RECONNAISSANCE VEHICLE

Country of Origin: Israel
Crew: 2 + 6
Engine: Dodge Model 225.2, 6-cylinder water cooled petrol engine, developing 120hp
Performance: Speed 100km/h (62.11mph) (road); range 550km (341.61mls) (road), 400km (248.45mls) (cross country); fuel 140 litres (30.84gal)
Armament: Four 7.62mm (0.3in) or 12.7mm (0.5in) MGs on individual mounts around the top of the hull
Armour: 10mm (0.40in) (max)

Dimensions: Length 5.023m (16ft 5in); width 2.03m (6ft 7in); height 1.66m (5ft 5in) (w/o armament)
Weight: 3,600kg (7,920lb) (empty)
History: A lightly armoured vehicle designed to replace the M551 as a reconnaissance vehicle in the Israeli Army. Designed from combat experience gained in the Middle East Wars, the RBY is in limited service with the Israeli ground forces and is also offered for export.

FIAT/OTO-MELARA TYPE 6616 ARMOURED CAR

Country of Origin: Italy

Crew: 3
Engine: Model 8062.24 super-charged in-line diesel, developing 160hp at 3,200rpm
Performance: Speed 100km/h (62.11mph) (road), 5km/h (3.11mph) (water); range 700km (434.78mls); fuel 150 litres (33.04gal)
Armament: 1 × 20mm (0.8in) Rh. 202 cannon, elevation +35°, depression -5° (400 rounds carried); 1 × 7.62mm (0.3in) coaxial MG (1,000 rounds); 2 × 3 smoke dischargers either side of turret

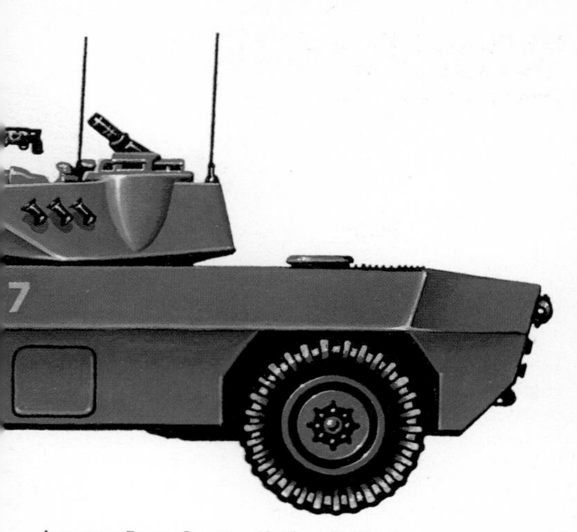

Armour: 5mm-8mm (0.2in - 0.32in)
Dimensions: Length 5.37m (17ft 7in); width 2.5m (8ft 2in); height 2.035m (6ft 8in) (top of turrret)
Weight: 7,400kg (16,280lb) (loaded), 6,900kg (15,180lb) (empty)
History: A private venture project by FIAT and OTO-MELARA, the Type 6616 has recently entered Italian service in limited numbers. The vehicle has also been offered for export and it is known that vehicles of this type are in use in Somalia and South Korea.

BRDM-2
ARMOURED RECONNAISSANCE VEHICLE

Country of Origin: USSR
Crew: 2 + 4 (depending on variant)

Engine: GAZ-41 V-8 water-cooled petrol engine, developing 140hp at 3,400rpm
Performance: Speed 100km/h (62.11mph) (road), 10km/h (6.21mph) (water); range 750km (465.84mls)
Armament: 14.5mm (0.57in) MG + 7.62mm (0.3in) MG
Weight: 7.0 tonnes (15,400 lb)
Armour: 10mm (0.40in)
Dimensions: Length 5.75m (18ft 10in); height 2.31m (7ft 6in); width 2.35m (7ft 8in)
History: The BRDM-2, which was introduced into Soviet service in 1966, represents a considerable advance over its predecessor the BRDM-1. It adopts the rear-engined arrangement more usual in reconnaissance vehicles, has greater speed and range, and also has a reasonable defensive armament. It has slightly improved ground clearance and amphibious

capability compared to the ARDM-1, and is also fitted with IR night-driving equipment. As well as filling the primary light reconnaissance role, this ubiquitous vehicle has been adapted for a wide range of other tasks. In the antitank role, the turret is removed; in the SAGGER variant there is a flat one-piece cover; and the AT-4/AT-5 version has a pedestal mounting. The command variant is also without turret but has a prominent radio antenna mounting on each side of the superstructure and often carries a box on the roof. The BRDM-2 is also used as the basis for the SA-9 GASKIN SAM system. The chemical reconnaissance version, the BRDM-2 RkH, is fitted with sampling equipment to detect the presence of chemical agents. On the rear deck are two contamination marker-flag dispensers.

BRDM-1
ARMOURED RECONNAISSANCE VEHICLE

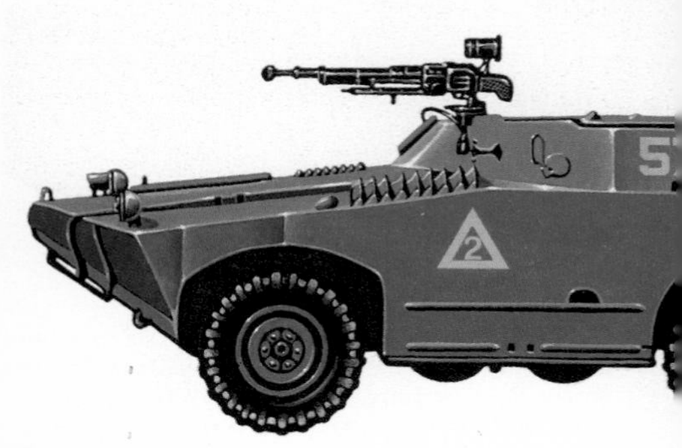

Country of Origin: USSR
Crew: 2 + 3 (depending on variant)
Engine: GAZ-40P 6-cylinder water-cooled petrol engine, developing 90hp at 3,400rpm
Performance: Speed 80km/h (49.69mph) (road), 9km/h (5.59mph) (water); range 500km (310.56mls)
Armament: 7.62mm (0.3in) MG
Weight: 5.6 tonnes (12,320 lb)
Armour: 10mm (0.40in)
Dimensions: Length 5.7m (18ft 8in); height 1.9m (6ft 2in); width 2.25m (7ft 4in)
History: Before 1959 the Soviet Army did not have a

purpose-built reconnaissance vehicle, employing instead modified wheeled APCs and motor cycles. The shortcomings inherent in these vehicles were largely overcome with the introduction of the BRDM (Bronevaya Rasvedyvakeinaya Dosernaya Maschina) armoured reconnaissance vehicle. It was relatively fast, possessed good cross-country performance due to the tyre pressurization system and retractable belly wheels, and, most importantly, it was amphibious. The only preparation required was to raise the bow trim blade and open the rear cover of the water-jet unit. In due course the basic vehicle was employed in other roles. Antitank GW versions carrying SNAPPER, SWATTER and SAGGER were introduced into motor rifle units, and command/OP and NBC reconnaissance versions were produced. It has for the most part been replaced by the BRDM-2 in first-line service, but is still to be found serving in large numbers in Soviet-aligned countries.

BA-64
LIGHT ARMOURED CAR

Country of Origin: USSR
Crew: 2
Engine: GAZ-MM, 4-cylinder water-cooled in-line petrol engine, developing 50hp at 2,800rpm

Performance: Speed 80km/h (49.69mph) (road), 9km/h (5.56mph) (water); range 500km (310.56mls)
Armament: 1 × 7.62mm (0.3in) SGMB MG (1,250 rounds carried) and/or 1 × 12.7mm (0.5in) DSHKM MG
Armour: 10mm (0.4in)
Dimensions: Length 5.70m (18ft 8in); width 2.25m (7ft 4in); height 1.90m (6ft 2in) (w/o armament)
Weight: 5,600kg (12,320lb) (loaded), 5,100kg (11,220lb) (empty)

History: This Second World War design was a lightly armoured development of the GAZ-63 'Jeep-type' vehicle, designed for liaison and reconnaissance. Extensively used by the Soviet Army until replaced in the 1950s, the BA-64 saw action in North Korea, which still uses the vehicle in front-line service. Known as 'Bobby' to the American Army who confronted it, the BA-64 is a vehicle of limited value, being easily destroyed. In Europe only the Albanian Army currently holds stocks.

FV601 ALVIS SALADIN
Mk2 ARMOURED CAR

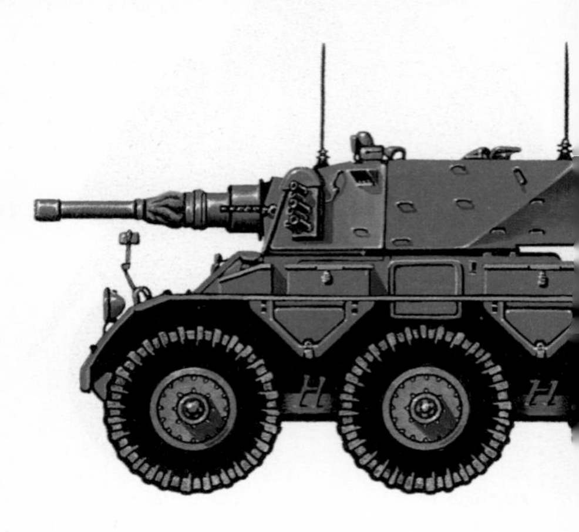

Country of Origin: United Kingdom
Crew: 3
Engine: Rolls-Royce B.80 Mk6A 8-cylinder petrol engine, developing 170hp at 3,750rpm
Performance: Speed 72km/h (44.72mph); range 400km (248.45mls)
Armament: 76mm (3in) gun; 7.62mm (0.30in) MG/(coaxial)

Weight: 11.59 tonnes (25,498lb)
Armour: 8.3mm (0.32in)
Dimensions: Length 5.284m (17ft 3in) (hull: 4.93m/16ft 2in); height 2.19m (7ft 2in); width 2.54m (8ft 3in)
History: The Saladin entered service in 1959 and remained in production until 1972. It is no longer in first-line service with the British Army, with the exception of 12 vehicles retained in Cyprus, although

a quantity remain in storage. It is still in widespread use in African and the Middle East countries. The 6×6 chassis, which is common to the Saracen APC and Stalwart high-mobility load carrier, provides a fast and versatile platform for the powerful ROF. 76mm (3in) gun. The vehicle's mobility was enhanced by fitting run-flat tyres, and it has a capacity to keep moving despite the loss of several wheels.

FV101 SCORPION
COMBAT VEHICLE RECONNAISSANCE (TRACKED)

Country of Origin: United Kingdom
Crew: 3
Engine: Jaguar 6-cylinder water-cooled petrol engine,

developing 190bhp at 4,750rpm
Performance: Speed 80.5km/h (50mph) (road); range 644km (400mls) (road); fuel 423 litres (93.17gal)
Armament: 1 × 76mm (3in) gun, elevation +35°, depression -10° (40 rounds carried); 1 × 7.62mm (0.3in) MG (3,000 rounds); 2 × 4-barrelled smoke dischargers
Dimensions: Length 4.794m (15ft 8in); width 2.235m (7ft 3in); height 2.102m (6ft 10in)
Weight: 8,000kg (17,600lb) (loaded)
Ground Pressure: 0.36kg/cm² (5.15psi)
History: Developed by Alvis in the United Kingdom,

the CVRT series has replaced all wheeled reconnaissance vehicles used by regular army units in Britain and BAOR. The Scorpion vehicle, properly described, is a scouting vehicle, the 76mm (3in) gun being used to get the vehicle 'out of trouble' rather than engage in offensive action. In British Army use, the vehicle is used with a mix of other CVRT family types, the 30mm (1.18in) Rarden cannon-equipped 'Scimitar', and the missile-equipped 'Striker'. The Scorpion has been widely exported; countries using it include Belgium, Ireland, Nigeria, Iran, Thailand, Spain and the Philippines.

FOX
COMBAT VEHICLE RECONNAISSANCE (WHEELED)

Country of Origin: United Kingdom
Crew: 3
Engine: Jaguar XK 4.2 (militarized) 6-cylinder, developing 190hp at 5,000rpm
Performance: Speed 104km/h (64.6mph) (road),

5km/h (water); range 434km (269mls) (road); fuel 145 litres (32.22gal)
Armament/Main Equipment: 30mm (1.2in) Rarden Cannon firing APDS, APSE and H.E. effective against APV's to 1000m (1,093yds) and SSVs to 2000m (2,187yds) + 7.62mm (0.3in) GPMG and smoke dischargers
Weight: 5,174kg (11,383lb) unladen; 6,356kg (13,983lb) battle laden
Dimensions: Length 4.22m (13ft 10in); width 2.13m (6ft 11in); height 2.2m (7ft 2in)
Ground Pressure: 0.46kg/cm^2 (6.57psi)
History: The Fox is a member of the latest family of armoured vehicles and the logical successor to the well-proven Ferret. Equipped with day and night sighting plus the battle-proven Rarden 30mm (1.18in), the Fox is capable of destroying all known

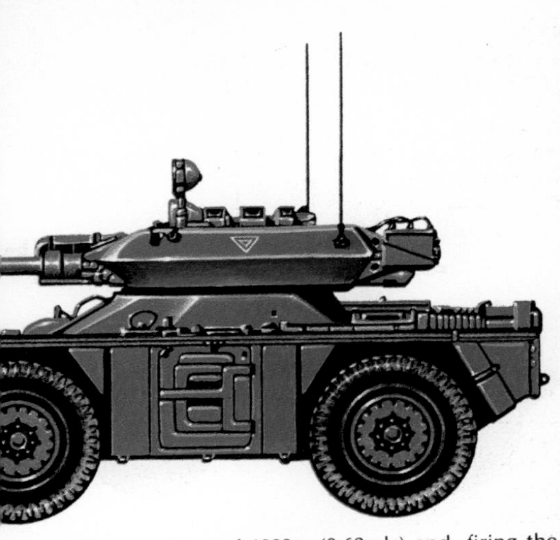

APCs at a range of 1000m (0.62mls) and, firing the special British APDS ammunition, can damage the side armour of an MBT at that range. Storage allows up to 99 rounds of 30mm (1.18in) and 2,600 rounds of 7.62mm (0.3in)to be carried. Light alloy armour gives protection against heavy machine gun fire and artillery splinters. The Fox is truly air-portable and three can be carried in a C-130 Hercules. It is also possible to parachute two together, using a special pallet.

FERRET Mk4
LIGHT SCOUT CAR

Country of Origin: United Kingdom

Crew: 2-3
Engine: Rolls-Royce B60 MK 6A 6-cylinder water-cooled petrol engine, developing 129bhp at 3,750rpm
Performance: Speed 80km/h (49.69mph) (road), 3.8km/h (2.36mph) (water); range 300km (186mls) (road); fuel 96 litres (21gal)
Armament: 0.30mm (1.18in) Browning MG
Ground Clearance: 0.43m (16.9in)

Dimensions: Length 4.095m (13ft 8in); width 2.13m (7ft 1in); height 2.336m (7ft 11in)
Weight: 5,400kg (11880lb) (loaded), 4,725kg (10,395lb) (empty)
History: The Ferret is probably one of the most widely distributed liaison/reconnaissance/IS patrol vehicles in the world, having been in production for nearly twenty years. Daimler, building on their World War II experience in this field, produced the first prototype Ferret in 1949. Delivery of production models commenced in 1952 and only ceased in 1971, when over 4,000 of various marks had been produced. The original Mk1 was a light liaison or reconnaissance vehicle, highly mobile, and with a very low silhouette. The armament was an optical Bren gun on a pintle mount. The Mk 1/2 featured a slightly raised superstructure with a folding lid. The Mk2 was fitted with a light, manually operated turret, mounting a 0.30 Browning machine gun. Subsequent variations to the basic Mk2 include a version with the turret mounted on a raised superstructure, which was used in the Far East, and the Mk 2/6 with a Vigilant antitank missile mounted on each side of the turret. The Mk4 is basically a rebuilt Mk2 but with larger wheels and tyres, stronger suspension, and disc brakes. The final development was the incorporation of a flotation screen, watertight storage compartments, and larger wheels. The Mk5 version, which had a special turret with four Swingfire missiles, is no longer in British Army service.

SHORLAND Mk3
ARMOURED PATROL CAR (WHEELED)

Country of Origin: United Kingdom
Crew: 3

Engine: Rover 6-cylinder petrol engine, developing 91bhp at 4,500rpm
Performance: Speed 88.5km/h (road); range 257km (standard tank), 514km (long range tank); fuel 64 litres (14.08gal) (standard), 128 litres (28.17gal) (long range)
Armament: 1 × 7.62mm (0.3in) machine gun (1,500 rounds carried); 2 × 4 smoke dischargers (optional)
Weight: 3.36 tonnes (7,392lb) (loaded); 2.93 tonnes (6,460lb) (empty)
Ground Pressure: 2.4kg/ cm² (34.2psi)
Dimensions: Length 4.59m (15ft); height 2.28m (7ft 5in); width 1.77m (5ft 9in)
Armour: 8.25mm-11mm (0.32in-0.43in)
History: The Landrover, in various forms, is in service with a considerable number of police and armed

forces throughout the world. The first Shorland armoured patrol car was based on a standard 2750mm (109in) LWB Landrover, with armoured bodywork by Short Brothers and Harland, of Belfast. The body is surmounted by a turret identical to that on the Mk2 and Mk4 Ferret armoured cars, and this turret can mount either a 7.62mm (0.3in) Browning machine gun or a 7.62mm (0.3in) GPMG. Smoke grenade dischargers and a variety of radio installations can be fitted. The Mk1 and Mk2 versions were powered by a Rover 4-cylinder petrol engine, and the latter has slightly improved armour. The Mk3, which was externally similar, had a six-cylinder engine. The latest version, the Mk4, has an extended bonnet, beneath which is the considerably more powerful Rover V-8 engine, introduced in 1980.

LYNX
COMMAND AND RECONNAISSANCE VEHICLE (TRACKED)

Country of Origin: USA

Crew: 3
Engine: GMC Detroit-Diesel 6V53 6-cylinder diesel, 215hp at 2,800rpm
Performance: Speed 70kph (43.48mph); range 523km (324.84mls)
Armament: 1 × 12.7mm (0.5in) machine gun, 1 × 7.62mm (0.30in) machine gun/25mm (1in) cannon; 2 × 3 smoke grenade launchers

Weight: 8.7 tonnes (19,140lb) loaded
Ground Pressure: 0.48kg/sq cm (6.84psi)
Dimensions: Length 4.59m (15ft); height 2.17m (7ft 1in); width 2.41m (7ft 10in)
History: In the early 1960s, after the widespread adoption of the M113 APC in NATO and Western aligned countries, the FMC Corporation developed a dedicated command and reconnaissance vehicle based on the M113. The US Army had already adopted the M114 (C + R), but a small quantity was ordered by the Netherlands Army as the M113 (C + R). The Canadian Army also purchased a quantity, in slightly altered form, as the Lynx Command and Reconnaissance Vehicle. In Canadian service the driver's position is at the front left-hand side, the radio operator, with a 7.62mm (0.3in) Browning machine gun pintle mounting, is behind the commander, with a cupola-mounted .50 caliber Browning M2 HB machine gun offset to the right between the driver and the radio operator. The Dutch version has the commander's cupola situated to the centre rear with the radio operator alongside the driver. Dutch vehicles have now been fitted with the Oelikon KBA-B 25mm (1in) cannon. The M113 aluminium armour and running gear have been retained, but it should be noted that the Lynx has only four road wheels. Some vehicles have a hatch on the right-hand side of the hull. It is sometimes colloquially referred to as 'M113'. FMC have also developed other armament options and engine and suspension improvements.

M-8
ARMOURED CAR

Country of Origin: USA
Crew: 4 (2-6)

Engine: Mercedes JXD 6-cylinder petrol, developing 110hp at 3,000rpm
Performance: Speed 90km/h (55.90mph); range 560km (347.83mls); fuel 212 litres (46.70gal)
Armament: 1 × 37mm (1.47in) gun; 1 × 7.62mm (0.3in) coaxial MG; 1 × 12.7mm (0.5in) A/A MG
Armour: 3mm-20mm (0.12in - 0.79in)
Dimensions: Length 5.003m (16ft 4in); width 2.54m (8ft 3in); height 2.25m (7ft 4in); ground clearance 0.29m (11.5in)
Weight: 7,892kg (17,362lb) (loaded)

History: The M-8 armoured car, known as the 'Greyhound' (due to its high forward speed) in British service, was one of the leading designs of the Second World War. Replaced very quickly in US service, who prefer tracked vehicles for the reconnaissance role, the M-8 was exported to many overseas countries. Today the vehicle remains in service with many Third World countries where it is used for internal security operations; however, it is very doubtful whether ammunition for its 37mm (1.45in) antitank gun remains in production anywhere in the world.

CADILLAC GAGE COMMANDO SCOUT

Country of Origin: USA
Crew: 1 + 1 (or 1 + 2)
Engine: Cummins V-6 diesel, developing 149hp at 3,300rpm

Performance: Speed 88.5km/h (54.97mph); range 800km (496.89mls); fuel 208 litres (45.81 gal)
Armament: 2 × 7.62mm (0.3in) MG
Dimensions: Length 4.699m (15ft 4in); width 2.057m (6ft 8in); height 2.235m (7ft 3in)
Weight: 6,577kg (14,469lb)
History: Developed privately by Cadillac Gage, the Commando Scout was first offered in 1977 but has not

yet entered production. Various armaments can be fitted, including twin 7.62mm (0.3in) MGs with 1,600 rounds ammunition, or 1 × 7.62mm (0.3in) and 1 × 12.7mm (0.5in) MG or twin 12.7mm (0.5in) MGs. A 20mm (0.79in) or 30mm (1.2in) cannon can also be fitted. Variants include an antitank version fitted with a Hughes TOW ATGW system, or an American M40 106mm (4.17in) recoilless rifle.

SAUREER 4K 4FA ARMOURED PERSONNEL CARRIER

Country of Origin: Austria
Crew: 2 + 8
Engine: Steyr Model 4FA 6-cylinder diesel, developing 250hp at 2,400rpm
Performance: Speed 65km/h (40.37mph) (road); range 370km (229.81mls); fuel 184 litres (40.53gal)
Armament: 1 × 20mm (0.8in) Oerlikon cannon with an elevation of +70° and a depression of −12°
Armour: 8mm-20mm (0.32in - 0.8in)
Dimensions: Length 5.40m (17ft 8in); width 2.50m (8ft 2in); height 2.17m (7ft 1in) (inc turret), 1.65m (5ft 4in) (hull top); ground clearance 0.42m (16.50in)

Weight: 15,000kg (33,000lb) (loaded)
Ground Pressure: 0.52kg/ cm^2 (7.44psi)
History: The first prototype, designated 3K 3H, was completed in 1958, powered by a 200hp Saurer diesel engine. The eventual 4K 4F entered production in 1961, with later models designated 4K 3FA and 4K 4FA. Production ceased in 1969 after 450 had been built. Later developments produced the 4K 7FA APC in 1970. Variants include the 4K 4FA-G1, armed with a 12.7mm (0.5in) M2 Browning MG. The vehicle is the standard APC of today's Austrian Army.

SIBMAS (6 × 6) ARMOURED PERSONNEL CARRIER (WHEELED)

Country of Origin: Belgium
Crew: 3 + 11
Engine: MAN D2566 MKF, 6-cylinder turbocharged diesel engine, developing 320hp at 1,900rpm
Performance: Speed 100kph (62.11mph) (4-11kph (2.5-6.83mph) in water); range 1,000km (621.12mls)
Armament: Various depending on role
Weight: 14.5 (31,900lb)-16.5 tonnes (36,300lb) loaded depending on configuration; 12,500kg (27,500lb) without
Dimensions: Length 7.32m (24ft); height 2.24m (7ft 4in) (2.77m (9ft 1in) with turret); width 2.5m (8ft 2in)

History: Designed with the non-European market in mind, the SIBMAS 6 × 6 amphibious APC/IS vehicle makes extensive use of MAN commercial vehicle components for ease of maintenance and spares availability, combined with proven reliability. The

first pre-production vehicle was revealed in 1979, three years after the completion of the prototype. The first customer was the Republic of Malaysia, which ordered 186 vehicles in 1981. The majority of these are the fire-support vehicle version, which features the Cockerill 90mm (3.54in) gun mounted in a turret produced by PRB s.a. Other options include single/twin 20mm (0.79in) cannon, the British 30mm (1.18in) Rarden cannon, and HOT or MILAN antitank missiles. Ambulance, command, ARV and mortar-carrier versions are also available.

BDX
ARMOURED PERSONNEL CARRIER (WHEELED)

Country of Origin: Belgium/Eire/United Kingdom
Crew: 2 + 10
Engine: Chrysler V-8, 180bhp petrol engine
Performance: Speed 100kph (62.11mph); range 500-900km (310.56-559mls)
Armament: Various
Weight: 10.7tonnes (2,3540lb)

Dimensions: Length 5.05m (16ft 6in); height 2.06m (6ft 9in) (2.84m (9ft 3in) with turret); width 2.5m (8ft 2in)
History: The BDX is the result of a development which started in the early 1970s as the Beherman-Timoney APC, named after the original Timoney design and

the associated Beherman-Demoen company. The original vehicle was a well-conceived basic wheeled APC affording a high degree of ballistic protection, making it particularly suited to the police/internal security role. The Belgian government purchased 183

for their police and air force. A variety of weapon fits is available, including 7.62mm (0.3in) and 12.7mm (0.5in) machine gun turrets, 20mm (0.79in) cannon and antitank missiles, and 81mm (3.2in) mortars.

COBRA
ARMOURED PERSONNEL CARRIER

Country of Origin: Belgium
Crew: 3 + 9
Engine: V-6 diesel, developing 143hp at 3,300rpm

Performance: Speed 80km/h (49.69mph) (road), 7km/h (4.35mph) (water); range 600km (372.67mls); fuel 260 litres (161.49gal)
Armament: 1 × 12.7mm (0.5in) MG; 2 × 101mm (4in) rocket launchers; 2 × 7.62mm (0.3in) bow MGs; 2 × 3 rifle grenade launchers
Dimensions: Length 4.2m (13ft 9in); width 2.7m (8ft 10in); height 1.65m (5ft 4in) (hull top); ground clearance 0.4m (15.75in)
Weight: 7,5000kg (16,500lb) (loaded)

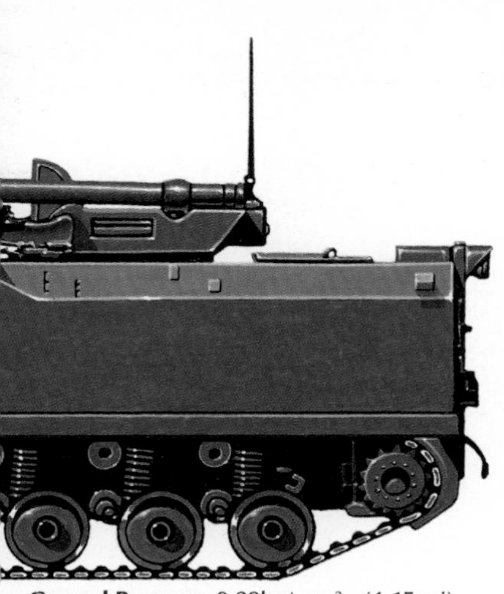

Ground Pressure: 0.29kg/ cm² (4.15psi)

History: Developed as a private venture, the Ateliers de Constructions Electriques de Charleroi produced two prototypes in 1980. The crew comprises two drivers at the front, each operating a bow-mounted 7.62mm (0.3in) MG, and a gunner/vehicle commander. There is a large door at the rear for the infantry to enter, but no facility for them to fire small arms once inside. Fully amphibious, the Cobra is propelled in the water by its tracks.

ENGESA EE-11 URUTU
ARMOURED PERSONNEL CARRIER

Country of Origin: Brazil
Crew: 14

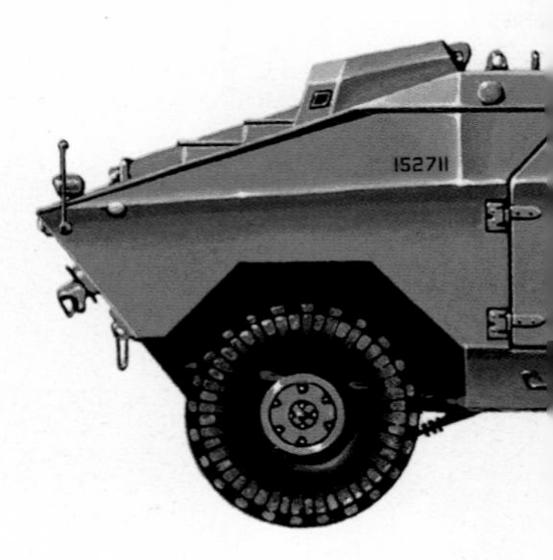

Engine: Mercedes-Benz OM-32 A 6-cylinder water-cooled turbocharged diesel, developing 190hp at 2,800rpm
Performance: Speed 90km/h (55.90mph) (road), 8km/h (4.97mph) (water); range 1,000km (621.12mls); fuel 380 litres (83.7gal)
Armament: 60mm (2.36in) breech-loaded Brandt mortar
Armour: 12mm (0.48in) (max)
Dimensions: Length 6m (19ft 8in); width 2.6m (8ft 6in);

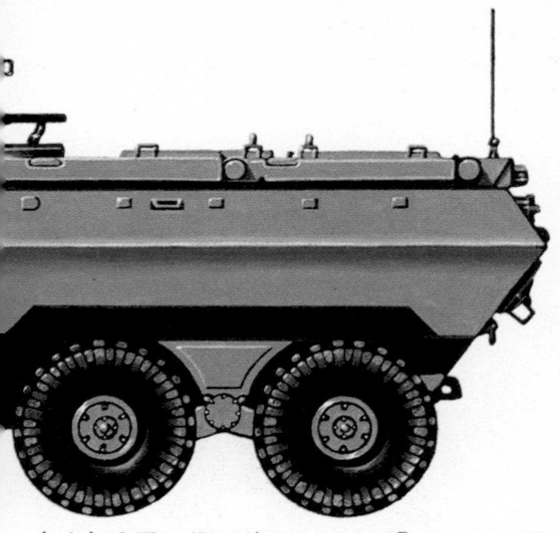

height 2.72m (8ft 11in) (top of MG mount); .09m (6ft 10in) (hull top)
Weight: 13,000kg (28,600lb) (loaded), 11,000kg (24,200lb) (empty)
History: The EE-11 Urutu was designed to meet the operational requirements of the Brazilian Army. A wide range of armament installation can be fitted, from simple pintle-mounted machine gun mounts to the tank-destroyer version equipped with a 90mm (3.54in) gun. This very cost-effective vehicle is in service with many South American countries and Libya.

OT-64
WHEELED ARMOURED PERSONNEL CARRIER

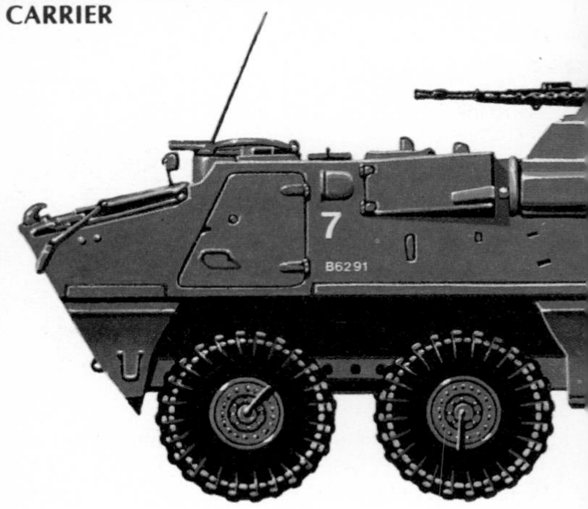

Country of Origin: Czechoslovakia
Crew: 2 + 14
Engine: Tatra T 928-14, V-8 diesel engine, developing 180hp at 2,000rpm
Performance: Speed 95kph (59mph) (9kph (5.59mph) in water); range 650km (403.73mls)
Armament: Various machine guns, depending on model

Weight: 14.4 tonnes (3,1680lb)
Dimensions: Length 7.44m (24ft 4in); height 2.3m (7ft 6in) (2.71m (8ft 10in) with turret); width 2.5m (8ft 2in)
History: The OT-64 series of APCs was developed by

Czechoslovakia in preference to the Soviet BTR-60 and entered Czech and Polish service in 1964. The basic model OT-64A was a basic armoured carrier, based on the Tatra 813 8 × 8 truck, and featured no additional armament except for the troops' personal weapons. The OT-64B has an open turret with either a 7.62mm machine gun or 12.7mm (0.5in) machine gun; this still serves with the Polish Army.

OT-62/TOPAS
ARMOURED PERSONNEL CARRIER (TRACKED)

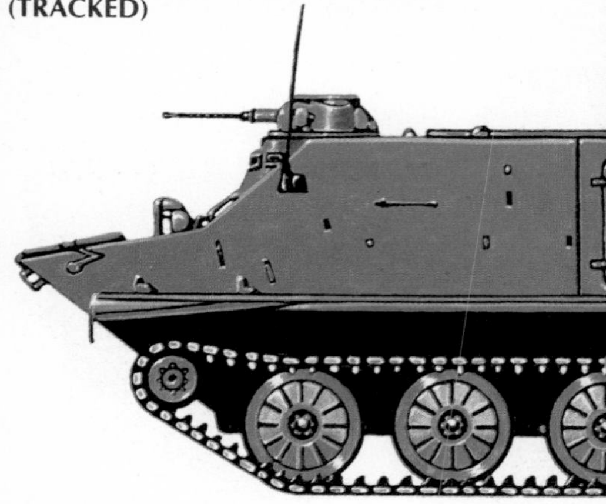

Country of Origin: Czechoslovakia
Crew: 2 + 14
Engine: Model PV-6, 6-cylinder, in-line diesel, developing 300hp at 1,800rpm
Performance: Speed 60kph (37.27mph) (11kph (6.83mph) in water); range 350-570km (217-354mls) (depending on model)
Armament: Various, depending on model
Weight: 16.4 tonnes (360,80lb)

Armour: 10-14mm (0.4-0.55in)
Dimensions: Length 7.0m (22ft 11in); height 2.10m (6ft 10in); width 3.14m (10ft 3in)
Ground Pressure: 0.53kg/ cm² (7.55psi)
History: Introduced into Czech service in 1964, the OT-62 is an improved version of the Soviet BTR 50 PK. It features a more powerful engine that gives superior performance despite the thicker armour and consequent weight increase. Access to the troop com-

partment is considerably eased by incorporating a large square hatch on each side of the hull. Another distinctive feature is the incorporation of a second cupola. The OT-62(A) is normally unarmed; the (B) and (C) models have a small turret on the right-hand cupola. The 62(B) turret is armed with a 7.62mm (0.3in) machine gun and an externally mounted T-21, 82mm (3.23in) recoilless gun (with a range of approximately 450m (500 yards)).

AMX-10P
ARMOURED PERSONNEL CARRIER

Country of Origin: France
Crew: 1 × 9
Engine: Hispano-Suiza HS 115-2, V-8 water-cooled diesel engine, developing 280hp at 3,800rpm
Performance: Speed 65kph (40.37mph) (road), 40kph

(24.84mph) (off road), 8kph (4.97mph) (water)
Armament: 1 × 20mm (0.8in); 1 × 7.62mm (0.3in) CA
Weight: 14.2 tonnes (31240lb) (combat)
Dimensions: Length 5.778m (18ft 11in); width 2.780m (9ft 1in); height 2.570m (8ft 5in)
History: The AMX-10P armoured personnel carrier is an amphibious, air-portable vehicle capable of good cross-country mobility. It is armed with a 20mm (0.79in) automatic gun plus a 7.62mm (0.3in) MG,

housed in a two-man turret located immediately to the rear of the driver's position, and above and forward of the troop compartment. Fitted with passive night-driving aids and good targeting optics, the AMX-10P carries a nine-man infantry section, capable of engaging light armoured vehicles while protected by an efficient NBC system. This vehicle is amphibious without preparation and driven either by water-jets or its tracks, or a combination of the two.

AMX VC-I
ARMOURED PERSONNEL CARRIER

Country of Origin: France
Crew: 3 + 10

Engine: SO FAM 8 Gbx 8-cylinder petrol engine, developing 250hp at 3,200rpm
Performance: Speed 65km/h (40.37mph) (road); range 350/400km (217/248mls) (road); fuel 410 litres (90.31gal)
Armament: 1 × 7.5mm (0.3in) or 7.62mm (0.3in) or 12.7mm (0.5in) MG
Armour: 10mm-30mm (0.4in - 1.19in)
Dimensions: Length 5.7m (18ft 8in); width 2.51m (8ft 2in); height 2.41m (7ft 10in) (with turret), 2.10m (6ft 10in) (hull top)

Weight: 15,000kg (33,000lb) (loaded), 12,500kg (27,500lb) (empty)
Ground Pressure: 0.70kg/ cm^2 (10.01psi)
History: A development of the highly successful AMX-13 light-tank design, the VCI has for many years been the major French armoured personnel carrier. Exported to many countries, particularly those already operating the AMX-13, the VCI exists in many versions including command post, ambulance, engineer, and mortar carrier, and an ammunition re-supply vehicle for self-propelled artillery batteries.

RENAULT VAB (4 × 4)
ARMOURED PERSONNEL CARRIER

Country of Origin: France
Crew: 2 + 10

Engine: MAN D 2356 HM 72 6-cylinder in-line, water-cooled diesel, developing 235hp at 2,200rpm
Performance: Speed 92km/h (57.14mph) (road), 7km/h (4.35mph) (water); range 1,000km (621.12mls); fuel 300 litres (66.08gal)
Armament: Creusot-Loire TLi 52A turret with a 7.62mm (0.3in) MG
Dimensions: Length 5.98m (19ft 7in); width 2.49m (8ft 2in); height 2.06m (6ft 9in) (w/o armament)

Weight: 13,000kg (28,600lb) (loaded), 11,000kg (24,200lb) (empty)
History: Rapidly becoming the major French APC, the VAB exists in both six-wheel and four-wheel versions. Fully amphibious, the VAB can carry differing armament fits depending on its role, the most common variant being the standard APC which carries a machine-gun turret. The VAB has been exported to many countries operating APVs of French origin.

BERLIET VXB-170
MULTI-ROLE VEHICLE

Country of Origin: France
Crew: 1 + 11
Engine: Berliet V8 diesel, developing 170hp at 3,000rpm

Performance: Speed 85km/h (52.8mph) (road), 4km/h (2.48mph) (water); range 750km (465.84mls); fuel 220 litres (48.46gal)
Armament: 1 × 90mm (3.54in) gun or 2 × 20mm (0.79in) cannon, 1 × 12.7mm (0.5in) MG or 2 × 7.62mm (0.3in) MGs, or 81mm (3.2in) mortar, plus various anti-tank missile systems and anti-aircraft missile systems
Armour: 7mm (0.28in) max
Dimensions: Length 5.99m (19ft 7in); width 2.50m (8ft

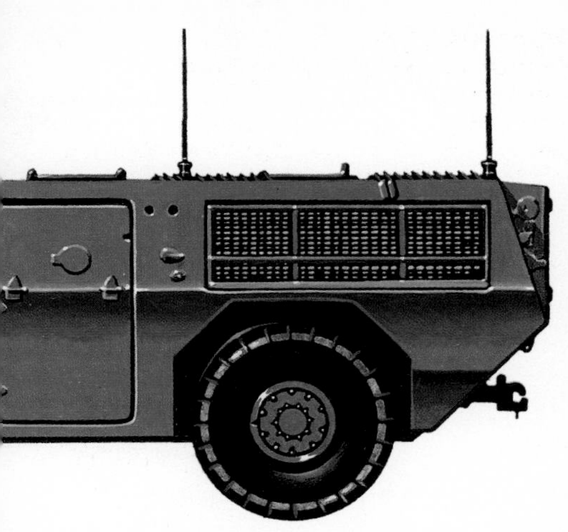

2in); height 2.05m (6ft 8in) (w/o turret)
Weight: 12,700kg (27,940lb) (loaded), 9,800kg (21,560lb) (empty)
History: The 4 × 4 VXB is of all-welded construction and is fully amphibious. Its crew can enter and leave the vehicle by way of side, rear and roof hatches. The VXB can be used in five basic roles; load carrier, light combat vehicle, reconnaissance vehicle, APC, and anti-riot vehicle. The VXB is used in the latter role by the French Gendarmerie.

PANHARD VCR
ARMOURED PERSONNEL CARRIER

Country of Origin: France
Crew: 3 + 9

Engine: Peugeot V-6 petrol, developing 140hp at 5,250rpm
Performance: Speed 110km/h (68.32mph) (road), 4.5km/h (2.8mph) (water); range 950km (590.06mls)
Armament: 1 × 7.62mm (0.3in), 12.7mm (0.5in) MG or 1 × 20mm (0.8in) cannon, or 1 × 60mm (2.38in) breech-loaded mortar dischargers either side of hull (optional) (front) and 1 × 7.62mm (0.3in) MG (rear) 2 smoke
Armour: 8mm – 12mm (0.32 – 0.48in)
Dimensions: Length 4.565m (14ft 11in); width 2.49m (8ft 2in); height 2.03m (6ft 7in) (commander's pos-

ition), 2.53m (8ft 3in) (inc armament)
Weight: 7,000kg (15,400lb)
History: The VCR (Véhicule de Combat à Roues - wheeled combat vehicle) range was developed by Panhard as a private venture aimed specifically at the export market. First shown in 1977, it entered production in 1978 equipped with the EWS missile UTM 800 turret. The basic APC carries ten infantrymen seated on benches, five along each side of the hull facing each other. The 4 × 4, also developed as a private venture, was announced in June 1979. It has identical automotive compounds to the 6 × 6.

PANHARD M3
ARMOURED PERSONNEL CARRIER

Country of Origin: France
Crew: 2 + 10
Engine: Panhard model 4 HD, 4-cylinder, air cooled

petrol engine, developing 90hp at 4,700rpm
Performance: Speed 100km/h (62.11mph) (road), 4km/h (2.48mph) (water); range 600km (372.67mls) (road); fuel 165 litres (36.34gal)
Armament: Creusot-Loire STB rotary support shield with 7.62mm (0.3in) MG
Armour: 8mm-12mm – (0.32in)-(0.48in)
Dimensions: Length 4.45m (14ft 7in); width 2.40m (7ft 10in); height 2.48m (8ft 1in) (turret), 2.0m (6ft 6in) (w/o turret)

Weight: 6,100kg (13,420lb) (loaded), 5,300kg (11,660lb) (empty)
History: Developed from the highly successful Panhard Armoured Car, the M3 Personnel Carrier has found a ready market in those countries that use the earlier vehicle. It is in service with the French Army and Gendarmerie and with many nations in Africa and South America. Versions of this vehicle have become the carrier and command post for the Crotale anti-aircraft missile system.

TRANSPORTPANZER 1
MULTI-PURPOSE ARMOURED VEHICLE
Country of Origin: West Germany
Crew: 2 + 10

Engine: 320hp Mercedes-Benz OM 402A, 8-cylinder water-cooled unloaded diesel (turbocharged)
Performance: Speed 105kph (65.22mph); range 800km (497mls); fuel 430 litres (95gal)
Armament: 1 × 20mm (0.8in) or 1 × 7.62mm (0.3in), 6 smaller dischargers
Weight: 17 tonnes (37,400lb) loaded, 14 tonnes (30,800lb) unloaded
Dimensions: Length 6.76m (22ft 2in); width 2.98m (9ft 9in); height 2.3m (7ft 6in)
History: Although manufactured by Thyssen Henschel, the original design was produced by

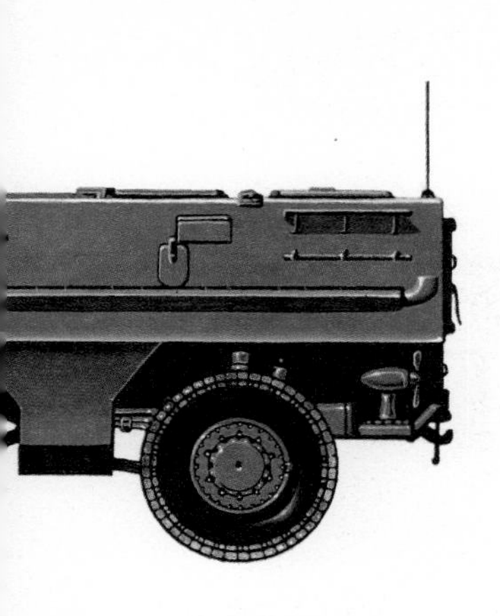

Daimler-Benz in response to a West German Army specification. A contract was awarded for 996 vehicles in 1977, with delivery commencing in 1979. It is known as 'Fuchs' (Fox) in German service. The TPz 1 is employed in a variety of roles by the Bundeswehr, including troop and cargo carrier, OP, EW, and NBC variant. Other versions proposed for export include ambulance, mortar, recce, ARV and maintenance roles. A 4 × 4 amphibious multipurpose vehicle has been ordered, and Krauss-Maffei have produced a private venture 8 × 8 version.

SCHUTZENPANZER SPZ 12-3
ARMOURED PERSONNEL CARRIER

Country of Origin: West Germany
Crew: 3 + 5

Engine: Rolls-Royce B81 80F, 8-cylinder petrol engine, developing 220hp at 4,000rpm

Performance: Speed 58km/h (36.02mph) (road); range 270km (167.70mls); fuel 340 litres (74.89gal)

Armament: 1 × 20mm (0.8in) Hispano-Suiza 820 gun, elevation +75°, depression -10° (2,000 rounds carried); 1 × 7.62mm (0.3in) MG (optional); 2 × 4 smoke grenade launchers

Armour: 8mm-30mm (0.32in - 1.2in)

Dimensions: Length 6.31m (20ft 8in) (inc gun), 5.56m (18ft 2in) (hull only); width 2.54m (8ft 3in); height 1.85m (6ft) (inc turret), 1.63m (5ft 4in) (w/o turret)
Weight: 14,600kg (32,120lb) (loaded)
Ground Pressure: 0.75kg/ cm^2 (10.73psi)

History: Designed originally in Switzerland by Hispano-Suiza, the SPz 12-3 APC was built in Great Britain as the standard infantry armoured vehicle of the reconstituted West German Army. Not liked in German service, the SPz 12-3 has been supplanted by the US-supplied M-113 and more recently by the Marder vehicle. Stocks of the SPz 12-3 are held by West Germany's territorial battalions who are tasked with the defence of rear areas.

SCHUTZENPANZER NEU MARDER
MECHANIZED INFANTRY COMBAT VEHICLE

Country of Origin: West Germany
Crew: 4 + 6

Engine: MTU MB 833 Ea-500, 6-cylinder diesel, developing 600hp at 2,200rpm
Performance: Speed 75km/h (46.58mph) (road); range 520km (322.98mls) (road); fuel 652 litres (143.61gal)
Armament: 1 × 20mm (0.8in) Rh 202 cannon, elevation +65°, depression -17° (1,1250 rounds carried); 2 × 7.62mm (0.3in) MG3 MGs – coaxial and at rear (5,000 rounds); 6 smoke dischargers on the turret
Dimensions: Length 6.79m (22ft 3in); width 3.24m (10ft 7in); height 2.95m (9ft 8in) (inc searchlight), 2.86m (9ft 4in) (turret top)

Weight: 28,200kg (62,040lb) (loaded)
Ground Pressure: 0.80kg/cm² (11.44psi)
History: A very important vehicle to the West German Army, the Marder is a highly capable infantry fighting vehicle, designed to accompany main battle tanks in extremely hostile environments. Standard to all West German armoured and mechanical divisions, the 'Marder' is a powerful vehicle, having its own anti-armour capability with its turret-mounted MILAN ATGW missile system and its double feed belt for the 20mm (0.79in) cannon.

HWK II
ARMOURED PERSONNEL CARRIER

Country of Origin: West Germany
Crew: 2 + 10
Engine: Chrysler 361 B, 8-cylinder petrol engine,

developing 211hp at 4,000rpm
Performance: Speed 65km/h (40.37mph) (road); range 320km (198.76mls) (road); fuel 300 litres (66.08gal)
Armament: 1 × 7.62mm (0.3in) MG or 1 × 12.7mm (0.5in) MG

Armour: 8mm-14.5mm (0.4in - 0.57in)
Dimensions: Length 5.05m (16ft 6in); width 2.53m (8ft 3in); height 1.585m (5ft 2in) (w/o MG)
Weight: 11,000kg (24,200lb) (loaded), 9,000kg (19,800lb) (empty)
Ground Pressure: 0.55kg/sq cm (7.87psi)

History: Developed as a private-venture candidate for a new APC to equip the West German Army in the early 1960s, the HWKII was unsuccessful. However, a small number went into production for the Mexican Army, where the type remains front-line equipment.

CONDOR
ARMOURED PERSONNEL CARRIER

Country of Origin: West Germany
Crew: 3 + 9

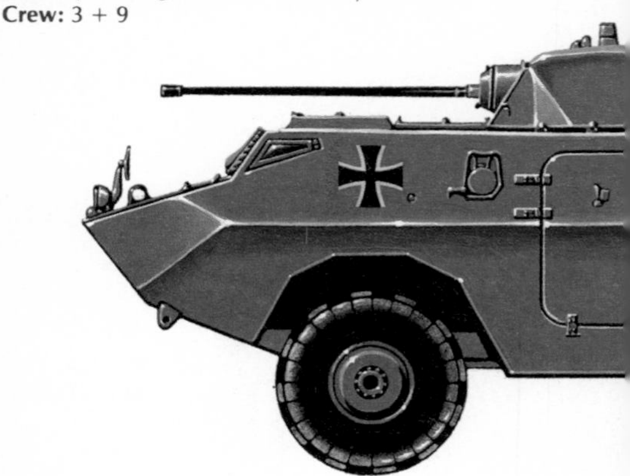

Engine: Daimler-Benz OM 352A 6-cylinder water-cooled diesel, developing 168hp
Performance: Speed 105km/h (65.22mph); range 500km (310.56mls) road; fuel 160 litres (99.38gal)
Armament: 1 × 20mm (0.8in) cannon, elevation +60°, depression -6°, 1 × 7.62mm (0.3in) coaxial MG (optional); 2 × 4 smoke dischargers (optional)
Dimensions: Length 6.06m (19ft 10in); width 2.47m (8ft

1in); height 2.79m (9ft 1in) (turret top); 2.10m (6ft 10in) (hull top)
Weight: 9,800kg (21,560lb) (loaded), 7,340kg

(16,148lb) (empty)
History: Designed to replace the UR-416, the Condor is fully amphibious. The basic vehicle can carry nine infantrymen in its APC role or can be converted into a specialist reconnaissance vehicle with a 20mm (0.79in) cannon, or a specialist tank destroyer version with MILAN or HOT ATGWS. Currently the vehicle is in service with the army of Ecuador.

UR-416
ARMOURED PERSONNEL CARRIER

Country of Origin: West Germany
Crew: 2 + 8

Engine: DB OM-352, 6-cylinder, water-cooled, in-line, diesel, developing 120hp at 2,800rpm
Performance: Speed 85km/h (52.8mhp) (road); range 700km (434.78mls) (road); fuel 150 litres (33.04gal)
Armament: 1 × 7.62mm (0.3in) MG, elevation +75°, depression -10°
Armour: 9mm (0.36in)
Dimensions: Length 4.99m (16ft 4in); width 2.30m (7ft 6in); height 2.25m (7ft 3in) (hull top)

Weight: 7,600kg (16,720lb) (loaded), 5,700kg (12,540lb) (empty)
History: A lightly armoured development of the famous Uni-Mog light truck, the UR-416 has been developed specifically for internal security and

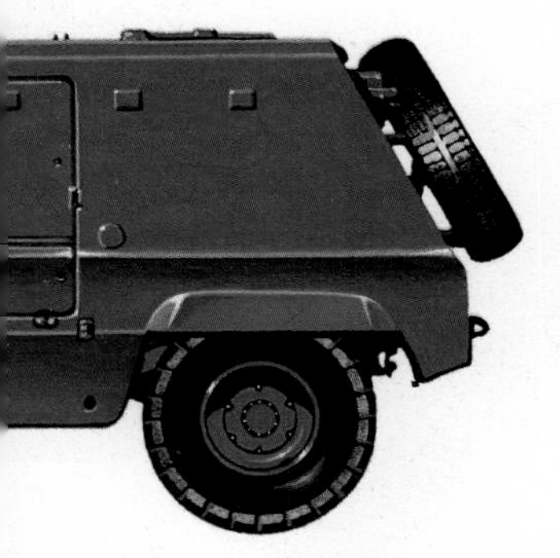

border patrol. Capable of being fitted with varying armaments from simple pintle-mounted machine gun mounts or power-operated turrets fitted with 20mm (0.79in) cannon, the vehicle is most commonly used as an APC. Found mostly in paramilitary usage rather than as army equipment, the UR-416 is used in considerable numbers by the West German Border Police and at many international airports around the world.

FIAT/OTO-MELARA TYPE 6614
ARMOURED PERSONNEL CARRIER

Country of Origin: Italy
Crew: 1 + 10

Engine: Model 8062.24 supercharged liquid-cooled in-line diesel, developing 160hp at 3,200rpm
Performance: Speed 100km/h (62.11mph) (road), 3.5km/h (2.17mph) (water); range 700km (434.78mls); fuel 142 litres (31.28gal)
Armament: 1 × 12.7mm (0.5in) MG
Armour: 6mm-8mm (0.24in-0.32in)
Dimensions: Length 5.86m (19ft 2in); width 2.5m (8ft

2in); height 1.75m (5ft 8in) (w/o armament)
Weight: 8,500kg (18,700lb) (loaded)
History: Typical of modern wheeled APCs, the 6614 can be fitted with a variety of differing weapon fits, ranging from rifle-calibre machine guns to the FIROS 48-round multi-rocket launcher. Used by spearhead units of the Italian Army, the vehicles are also in service in South Korea, Peru, Somalia and Tunisia.

TYPE 73
ARMOURED PERSONNEL CARRIER

Country of Origin: Japan

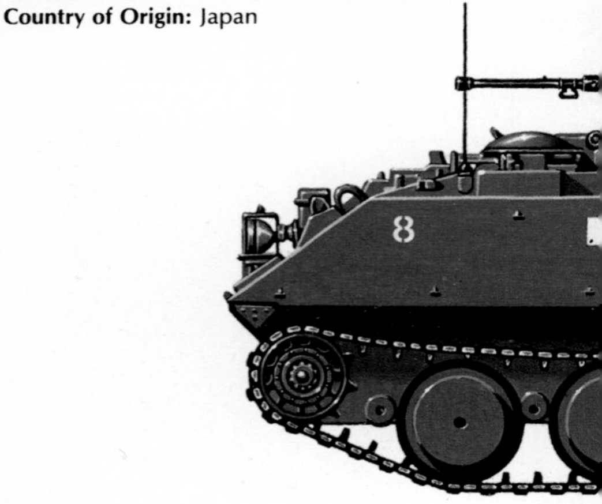

Crew: 3 + 9

Engine: Mitsubishi model 42F 4-cylinder air-cooled diesel, developing 300hp at 2,200rpm

Performance: Speed 70km/h (43.48mph) (road), 7km/h (4.35mph) (water); range 300km (186.34mls), fuel 450 litres (99.12gal)

Armament: 1 × 12.7mm (0.5in) roof-mounted MG, elevation +60°, depression -10°; 1 × 7.62mm (0.3in) bow MG; 2 × 3 smoke dischargers

Dimensions: Length 5.8m (19ft); width 2.8m (9ft 2in); height 2.2m (7ft 2in) (with 12.7mm (0.5in) MG), 1.7m (5ft 6in) (hull top)
Weight: 13,300kg (29,260lb) (loaded)
History: The Type 73 is in full production for the Japanese Ground Self-Defence Force as its standard armoured personnel carrier. The vehicle is fully amphibious and can carry its crew in full NBC conditions.

TYPE SU60
ARMOURED PERSONNEL CARRIER

Country of Origin: Japan
Crew: 4+6

Engine: Mitsubishi 8 HA-21 WT, V-8 air-cooled turbo-charged diesel, developing 220hp at 2,400rpm
Performance: Speed 45km/h (27.95mph); range 230km (142.86mls)
Armament: 1 × 12.7mm (0.5in) M2 MG; on roof 1 × 7.62mm (0.3in) M1919A4 bow MG
Dimensions: Length 4.85m (15ft 10in); width 2.40m (7ft 10in); height 2.31m (7ft 6in) (inc MG), 1.70m (5ft 6in) (w/o MG)
Weight: 11,800kg (25,960lb) (loaded), 10,600kg (23,320lb) (empty)
Ground Pressure: 0.57kg/sq cm (8.15psi)
History: Designed specifically for the Japanese Self-Defence Forces, some 400 of these unique vehicles

were produced between 1960 and 1970. A simplistic 'battle taxi', the SU60 has no amphibious capacity or sophisticated night-vision aids. Standard equipment in Japanese mechanized battalions, the SU60 is configured to the dimensions of the average Japanese, which makes it somewhat restricted for use by other nations. This point is further underlined by the fact that the US M-113, which is the most widely used Western APC, has never been used by the Japanese as, in a nation where the average height of males of military age is 1.625m (5ft 4in), the M-113 appears huge. The SU60 is otherwise a vehicle of absolutely no tactical value whatsoever.

YP 408
WHEELED APC

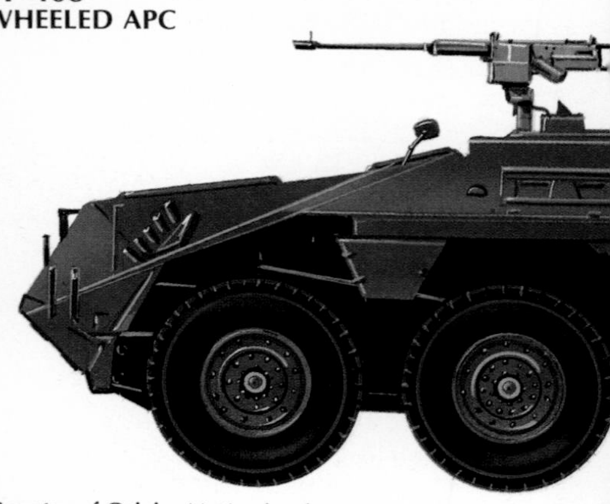

Country of Origin: Netherlands
Crew: 2+10
Engine: DAF D5575, 6-cylinder in-line, water-cooled, turbo charged diesel engine, developing 165hp at 2,400rpm
Performance: Speed 80km/h (49.69mph); range 500km (310.56mls)
Armament: 1 × 12.7mm (0.5in) machine gun; 2 × 3 barrelled smoke dischargers
Weight: 12.0 tonnes (26,400lb)
Armour: 15mm (0.6in) max

Dimensions: Length 6.32m (20ft 5in); height 1.18m (3ft 10in) (hull top); width 2.4m (7ft 10in)

History: The YP 408 has been the standard infantry APC in Netherlands Army service since 1964. The first prototype was produced by DAF in 1958, based substantially on the DAF YA 326 3-tonne 6×6 truck. This is reflected in the fact that only the front and rear two axles are driven on the YP 408. Between 1964 and 1968 some 750 vehicles were delivered to the Netherlands Army. As well as the basic infantry section vehicle, there are command versions, a 120mm (4.72in) mortar-towing vehicle a load carrier, and a surveillance vehicle equipped with the British ZB 298 radar.

CHAIMITE
ARMOURED PERSONNEL CARRIER

Country of Origin: Portugal
Crew: 11

Engine: V-8 petrol, developing 210hp at 4,000rpm, or V6 diesel
Performance: Speed 99km/h (61.49mph) (road), 7km/h (4.35mph) (water); range 804-965km (500-600mls) (road, petrol engine model), 1,367-1,529km (road, diesel engine model); fuel 300 litres (66gal)
Armament: Fitted with BRAVIA turret twin 7.62mm (0.3in), twin 5.56mm (0.22in) or one 7.62mm (0.3in) and one 12.7mm (0.5in) MG
Armour: 6.35mm-9.35mm (0.25in - 0.37in)

Dimensions: Length 5.606m (18ft 4in); width 2.26m (7ft 4in); height 2.26m (7ft 4in) (turret top), 1.84m (6ft) (hull top)
Weight: 7,300kg (16,060lb) (loaded)
History: This vehicle is a licensed variant of the US-designed Commando armoured car and APC. Following the withdrawal of Portugal from its overseas colonies, the Portugese Army had only very old equipment. The Chaimite is the first armoured vehicle to be manufactured in Portugal.

RATEL 20
INFANTRY FIGHTING VEHICLE

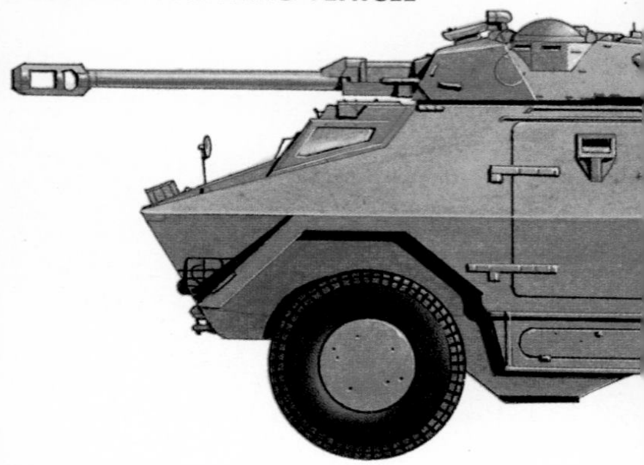

Country of Origin: South Africa
Crew: 3+7
Engine: 6-cylinder turbocharged diesel
Performance: Speed 105km/h (65.22mph); fuel 480 litres (105.73gal)
Armament: 1 × 20mm (0.8in) cannon (500 rounds carried); 2 × 7.62mm (0.3in) MGs – coaxial and A/A (500 rounds); 2 smoke dischargers either side of turret
Dimensions: Length 7.21m (23ft 7in); width 2.7m (8ft 10in); height 3.11m (10ft 2in) (overall), 2.30m (7ft 6in) (hull top)

Weight: 17,000kg (37,400lb) (loaded); 15,000kg (33,000lb) (unloaded)
History: Designed and manufactured in South Africa, the RATEL infantry fighting vehicle bears a close family resemblance to the Panhard series of armoured cars which have been license-produced under the name ELAND. The RATEL has seen action in South Africa's border conflicts, particularly in Angola, and has been exported in limited quantities to Morocco.

BMR-600
INFANTRY FIGHTING VEHICLE

Country of Origin: Spain
Crew: 2 + 16

Engine: Pegaso 9157/8 diesel, 306hp at 2,600rpm
Performance: Speed 100km/h (62.11mph) (road), 10km/h (6.21mph) (water); range 900km (559mls) fuel 320 litres (70.48gal)
Armament: 1 × 7.62mm (0.3in) MG (2.500 rounds carried)
Dimensions: Length 6.15m (20ft 2in); width 2.49m (8ft 2in); height 2.00m (6ft 6in) (hull top), 2.36m (7ft 8in)

(inc armament)
Weight: 13,000kg (28,600lb) (loaded)
History: The Blindado Medio de Ruedas 600 was developed from the early 1970s to meet the require-

ments of the Spanish Army, which until this time had relied upon obsolescent US equipment. The vehicle is fully amphibious, being propelled in the water by two DOWTY water jets mounted on the rear of the engine deck. An unusual feature of the BMR-600 is its hydropneumatic suspension, which allows the driver to adjust the vehicle's ground clearance to suit the ground conditions.

PBV 302
ARMOURED PERSONNEL CARRIER
Country of Origin: Sweden
Crew: 2 + 10

Engine: Volvo THD 100B, 6-cylinder, in-line, turbo-charged diesel, developing 280hp at 2,200rpm
Performance: Speed 66km/h (40.99mph) (road), 8km/h (4.97mph) (water); range 300km (186.34mls) (road); fuel 285 litres (62.78gal)
Armament: 1 × 20mm (0.8in) cannon, elevation +50°, depression -10° (505 rounds carried); 2 × 4 smoke grenade launchers
Dimensions: Length 5.35m (17ft 6in); width 2.86m (9ft 4in); height 2.50m (8ft 2in) (inc turret), 2.06m (6ft 9in) (hull top)

Weight: 13,500kg (29,700lb) (loaded)
Ground Pressure: 0.60kg/ cm^2 (8.58psi)
History: Entering service with the Swedish Army in 1966, the Pansarbandvagn 302 is the standard armoured personnel carrier of this force. Fully amphibious, the vehicle is equipped with a powerful Hispano Suiza cannon for use against both ground and airborne targets. Many variants of the basic vehicle exist, including a command vehicle, an armoured observation vehicle, and a recovery vehicle.

BMP-1
MECHANIZED INFANTRY COMBAT VEHICLE

Country of Origin: USSR
Crew: 3+8

Engine: 300bhp D, 6-cylinder, water-cooled
Performance: Speed (road) 80km/h (49.69mph), (water) 8km/h (4.97mph); range 500km (310.56mls); fuel 460 litres (101gal)
Armament: 1 × 73mm (2.89in); 1 × 7.62mm (0.3in) coax; 1 × 'SAGGER' ATGW launcher
Weight: 13,500kg (29,700lb) (combat), 12,500kg (27,500lb) empty
Armour: 23mm (0.9in) max.
Dimensions: Length 6.74m (22ft 1in); height 2.15m

(7ft); width 2.94m (9ft 7in)

Ground Pressure: 0.57kg/ cm^2 (8.12psi)

History: The BMP-1 is gradually replacing the other types of APC used in motor rifle and tank divisions in Soviet and Warsaw Pact armies. Armed with the 73mm (2.89in) 2A28 smooth-bore gun housed in the same one-man turret as used on the BMD, this well-armoured amphibious vehicle has many sophisticated assets not normally found on Soviet vehicles. The crew has good visibility, and improved targeting aids.

BMP
AIRBORNE COMBAT VEHICLE
Country of Origin: USSR
Crew: 7

Engine: V-6 liquid cooled diesel, developing 290hp
Performance: Speed 80km/h (49.69mph) (road), 10km/h (6.21mph) (water); range 320km (198.76mls); fuel 300 litres (66.08gal)
Armament: 1 × 73mm (2.89in) gun, elevation +33°, depression -4° (40 rounds carried); 3 × 7.62mm (0.3in) PKT MGs – coaxial and two bow (2,000 rounds for coaxial); 1 × 'sagger' ATGW launcher (3 missiles)
Armour: 6mm-25mm (0.24in - 0.99in)
Dimensions: Length 5.41m (17ft 8in); width 2.55m (8ft 4in); height 1.77m (5ft 9in)
Weight: 8,000kg (17,600lb) (loaded)

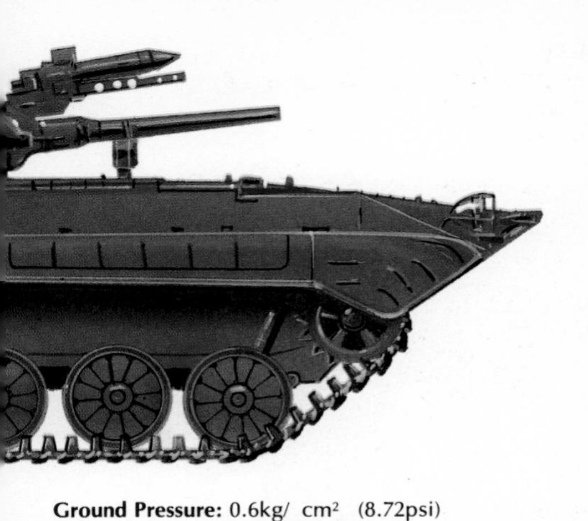

Ground Pressure: 0.6kg/ cm² (8.72psi)
History: A highly innovative vehicle, the BMD equips Soviet airborne divisions. Designed as an airportable and air-droppable armoured support vehicle, the BMD offers protection from small-arms fire to its crew and passengers. Equipped with the low pressure 73mm (2.87in) gun common to the BMP-1 APC, and the SAGGER antitank missile, the vehicle offers airborne troops a valuable fire-support facility. Recent developments of the BMD is a version with a 30mm (1.2in) cannon, and an unarmed command vehicle.

BTR-50 P
ARMOURED PERSONNEL CARRIER (TRACKED)

Country of Origin: USSR
Crew: 2+20
Armament: 1 × 7.62mm (0.3in) MG

Engine: Model V-6, 6-cylinder, in-line, water-cooled diesel, developing 240hp at 1,800rpm
Performance: Speed 44km/h (27.33mph) (road) 11km/h (6.83mph) (water); range 400km (248.45mls); fuel 400 litres (88gal)
Weight: 14,200kg (31,240lb)
Armour: 10-14mm (0.4in - 0.55in)
Dimensions: Length 7.08m (23ft 2in); height 1.97m (6ft 5in); width 3.14m (10ft 3in)
Ground Pressure: 0.51kg/cm² (7.27psi)

History: The BTR-50 P series APC was introduced in the late 1950s and was for many years the mainstay of the Soviet-motor rifle regiments until its replacement by the BMP-1. Designed around the chassis of the PT-76 light amphibious tank, and the basis for the Czechoslovakian OT-62, the BTR-50 P has a forward crew compartment, with an open-topped troop compartment behind, capable of carrying up to 20 infantry on bench seats that run across the width of the vehicle.

MCV-80 WARRIOR
MECHANICAL INFANTRY COMBAT VEHICLE

Country of Origin: United Kingdom
Crew: 10
Engine: Rolls-Royce CV8 TCE-8 diesel, developing 800bhp at 2,300rpm

Performance: Speed 75km/h (46.6mph) max.; range 500km (310mls)
Armament: 1 x 30mm (1.2in) Rarden cannon, 1 x 7.62mm (0.3in) coaxial MG, 2 x 4 smoke dischargers
Dimensions: Length 5.42m (17ft 8in); width 2.8m (9ft 2in), height 2.82m (9ft 3in) inc turret, ground clearance 0.5m (1ft 7in)
Weight: 20,000kg (44,000lb)
History: The GKN Sankey-developed MCV-80 (Mechanical Combat Vehicle 80), now named 'Warrior', is

seen as the eventual replacement for the FV432. Having only become available in 1986, the British Army has a need for some 2000 vehicles of this type. The

driver is seated front left, with the engine compartment to the right. The two-man turret is central, with the troop compartment at the rear. The infantry are unable to use their small arms from within. The MCV-80 carries a full range of passive night-vision equipment. Planned variants include a platoon vehicle, a command vehicle, mortar vehicle, artillery command post, all with GPMG turrets, and an engineer combat vehicle with a GPMG turret and an EMI Ranger antipersonnel mine-laying system.

ALVIS STORMER
ARMOURED PERSONNEL CARRIER

Country of Origin: United Kingdom
Crew: 3+8

Engine: Perkins T6/3544 6-cylinder turbocharged diesel, developing 200bhp at 2,600rpm
Performance: Speed 72km/h (44.72mph) road, 6.5km/h (4.04mph) water; range 800km (497mls) road; fuel 405 litres (89.2gals)
Armament: 1 × 20mm (0.8in) or 30mm (1.2in) cannon; 1 × 76mm (3in) or 90mm (3.56in) gun, twin 20mm (0.8in) A/A guns
Dimensions: Length 5.3m (17ft 4in); width 2.374m (7ft 8in), 2.654m (8ft 7in) with applique armour; height 2.374m (7ft 8in); ground clearance 0.362m (1ft 2in)
Weight: 10,689kg (23,516lb) loaded, 8,740 kg (19,228lb) empty

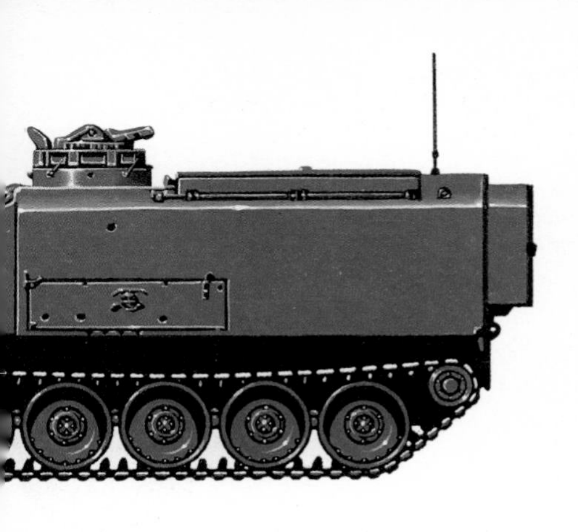

Ground Pressure: 0.37kg/ cm^2 (5.29psi)
History: A full-tracked prototype APC was built by the Royal Ordnance Factory, Leeds, in the late 1970s, designation FV4333. Now in production, the Stormer was developed from the Spartan APC and incorporates a longer hull which is also very slightly wider. Alvis purchased the manufacturing rights in 1980 and named the vehicle Stormer in 1981. The eighth member of the Alvis Scorpion family, the Stormer is an all-welded aluminium armour construction. It is fully amphibious, and is fitted with a full range of night-vision equipment.

FV603 ALVIS SARACEN ARMOURED PERSONNEL CARRIER

Country of Origin: United Kingdom
Crew: 2+10
Engine: Rolls-Royce B80 Mk 6A, 8-cylinder petrol engine, developing 160hp at 3,750rpm
Performance: Speed 72km/h (44.7mph) road; range

400km (248.5mls); fuel 200 litres (44gals)
Armament: 2 × 7.62mm (0.3in) MGs – one in turret, elevation +45°, depression –15°, one on ring mount at rear (3000 rounds carried), 2 × 3-barrelled smoke dischargers
Armour: 8mm-16mm (0.32in-0.64in)
Dimensions: Length 5.233m (17ft 2in) overall; width 2.539m (8ft 4in); height 2.463m (8ft 1in); ground clearance 0.432m (1ft 5in)
Weight: 10,170kg (22,374lb) loaded; 8,640kg (19,008lb) empty
Ground Pressure: 0.98 kg/cm² (14psi)

History: The FV603 Saracen APC was designed in the late 1940s, and the first prototype appeared early in 1952. Making full advantage of shared components in common with the Saladin armoured car and FV622 Stalwart high-mobility load carrier, the Saracen was in production over a period of twenty years from 1952 to 1972, eventually being replaced in British Army service with the tracked FV432 APC. Variants include the FV603 (C) with reverse-flow cooling for operation in hot climates and the FV604 Command Post without turret. The FV610 Command Post was similar to the FV604 but with height increased to 2.36m (7ft 9in).

DRAGOON 300
MULTI-MISSION VEHICLE

Country of Origin: USA

Crew: 9
Engine: Detroit-Diesel 6V53T, 6-cylinder turbocharged diesel, developing 300bhp at 2,100 rpm
Performance: Speed 116km/h (72mph) road; 5.5km/h (3.5mph) water; range 1,045km (649mls); fuel 341 litres (75.2gals)
Armament: Turret/pintle-mounted 7.62mm (0.3in) or 12.7mm (0.5in) MGs, turret-mounted 20mm (0.8in) or 25mm (1in) cannon, turret-mounted 90mm (3.6in) gun (Cockerill or Mecar) and ATGWS
Dimensions: Length 5.588m (18ft 3in); width 2.438m (8ft); height 2.642m (8.67ft) turret, 2.133m (7.0) hull; ground clearance 0.692m (2.3ft) hull
Weight: 11,830kg (26,026lb) loaded, 9,072kg (19,958lb)

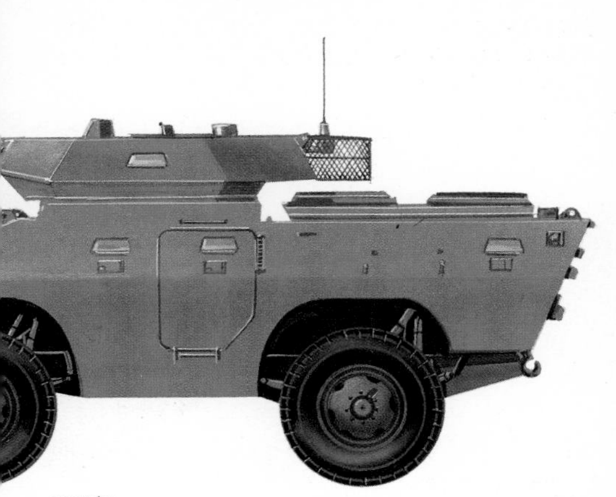

empty

History: Designed by the Verne Corporation, the prototype was produced in 1978 and was followed by 17 pre-production vehicles made by the Dominion Manufacturing Company and marketed by the Arrowpointe Corporation. Now in production, the Dragoon 300 family of vehicles relies to a large degree on commonality with the M-113A2 APC – including the engine – and the M-809 (6×6) 5-ton truck for axles, suspension, brakes and steering. The Dragoon 300 is fully amphibious and is propelled through the water by its wheels.

CADILLAC GAGE COMMANDO V-300 MULTI-MISSION VEHICLE

Country of Origin: USA
Crew: 3+9

Engine: V-8-555 diesel, developing 250bhp at 3,000rpm
Performance: Speed 88.51km/h (55mph) road, 4.8km/h (3mph) water; range 644 km (400 mls); fuel 284 litres (62.56gals)
Armament: One-man turret with twin 7.62mm (0.3in) MGs; or one 7.62mm (0.3in) and one 12.7mm (0.5in) MGs; or one 20mm (0.8in) cannon and one coaxial 7.62mm (0.3in) MG; or two-man turret with one 76mm (3in) gun and one coaxial 7.62mm (0.3in) MG; or one 90mm (3.56in) gun and coaxial 7.62mm (0.3in)

MG

Dimensions: Length 6.4m (20ft 11in); width 2.54m (8ft 3in); height 1.981m (6ft 5in) hull top, 2.692m (8ft 9in) with 90mm (3.56in) turret; ground clearance 0.533m (1ft 8in) max

Weight: 12,700kg (27,940lb) max

History: Now in production, the Cadillac Gage Commando V-300 (6×6) is available with a variety of armament installations, including the Emerson turret as fitted to the M-901 improved TOW Vehicle with two Hughes TOW ATGW in the ready-to-launch position.

CADILLAC GAGE COMMANDO V-150
MULTI-MISSION VEHICLE
Country of Origin: USA

Crew: 12 including commander, gunner and driver
Engine: 202hp diesel
Performance: Speed 88km/h (54.66mph) road; 4.8km/h (2.98mph) water; range 643km (399mls); fuel 303 litres (67gals)
Armament: Pintle-mounted 7.62mm (0.3in) MG
Dimensions: Length 5.689m (18ft 7in); width 2.26m (7ft 4in); height 1.981m (6ft 5in) over hull; ground clearance 0.381m (1ft 2in)

Weight: 9,888kg (21,754lb)
History: Replacing both the V-100 and V-200, the widely used V-150 entered production in 1971 and was the forerunner of the V-300. All these vehicles were made by the Cadillac Gage Company. Numerous variants are available in addition to the APC detailed above, and include a one-man turret armed with twin 7.62mm (0.3in) or one 7.62mm (0.3in) and one 12.7mm (0.5in) MG.

INFANTRY FIGHTING VEHICLE/CAVALRY FIGHTING VEHICLE

Country of Origin: USA
Crew: 3+6

Engine: Cummins VTA-903 8-cylinder turbocharged diesel, developing 500bhp at 2,400 rpm
Performance: Speed 66km/h (41mph) road; 7.2km/h (4.5mph) water; range 483km (300mls); fuel 662 litres (146gals)
Armament: 1 × 25mm (1in) cannon, elevation +60°, depression -10° (900 rounds carried), 1 × 7.62mm (0.3in) coaxial MG (4,400 rounds carried), twin launcher for Hughes TOW ATGW (7 TOWs), 2 × 4 smoke dischargers
Dimensions: Length 6.453m (21ft 1in); width 3.2m (10ft 5in); height 2.972m (9ft 8in) overall, 2.565m (8ft 4in) turret roof; ground clearance 0.457m (1ft 5in)

Weight: 22,045kg (48,500lb) loaded; 18,869kg (41,512lb) empty

History: Prototypes designated XM723 were built between 1973 and 1975 under contract by the FMC Corporation, and the project developed into the Fighting Vehicle System (FVS) with a two-man turret. The Infantry Fighting Vehicle was designated XM2 and the Cavalry Fighting Vehicle XM3, later standardized as the M2 and M3, which both became available in 1981. Both vehicles are fully amphibious, being propelled in the water by their tracks. They are fitted with an NBC system and passive night-vision equipment as standard.

ARMOURED INFANTRY FIGHTING VEHICLE

Country of Origin: USA
Crew: 3 + 7

Engine: Detroit Diesel 6V53T, V6 turbocharged diesel, developing 264bhp at 2,800rpm
Performance: Speed 61km/h (37.89mph) road, 6.3km/h (3.91mph) water; range 490km (304mls); fuel 416 litres (92gals)
Armament: 1 × 25mm (1in) cannon, elevation +50°, depression -10° (324 rounds carried), 1 × 7.62mm (0.3in) coaxial MG (1,840 rounds carried), 6 smoke dischargers
Armour: Aluminium and steel
Dimensions: Length 5.258m (17ft 2in); width 2.819m (9ft 2in); height 2.794m (9ft 1in) inc turret, 1.854m (6ft) front, hull top; ground clearance 0.432m (1ft 5in)

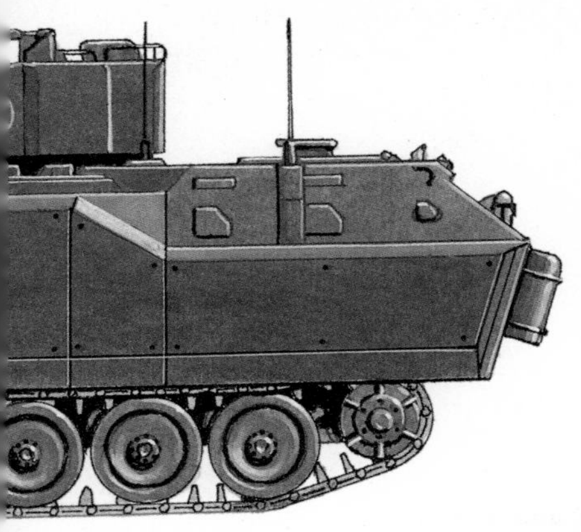

Weight: 13,687kg (30,111lb) loaded, 11,405kg (25,091lb) empty

History: The AIFV is in service with Belgium, Netherlands and the Philippines, and has a full range of night-vision equipment as standard. A fully amphibious vehicle, it is propelled in the water by its tracks, and has a superb cross-country performance. The United States Army originally ordered two vehicles designated XM765 for use in Korea from the FMC Corporation. Further developments led to the much-improved AIFV which appeared in prototype form in 1970.

M-59 APC

Country of Origin: USA
Crew: 2+10
Engine: 2 × GMC Model 302, 6-cylinder, water-cooled, in-line, petrol engines, each developing 127hp at 3,350rpm
Performance: Speed 551.50km/h (32mph) road, 6.9km/h (4.3mph) water; range 164km (102mls) road; fuel 518 litres (114gals)
Armament: 1 × 12.7mm (0.5in) M2MG (2,205 rounds carried)
Armour: 16mm (0.63in)
Dimensions: Length 5.613m (18ft 4in); width 3.263m (10ft 8in); height 2.768m (9ft) inc cupola, 2.387m (7ft

9in) hull top; ground clearance 0.457m (1ft 5in)
Weight: 19,323kg (42,511lb) loaded, 17,916kg (39,415lb) empty
Ground Pressure: 0.51kg/sq cm (7.29psi)
History: The M-59 has been replaced in the United States Army by the M-113, whilst still in service in Greece and Turkey. Originally designed in 1951 by the FMC Corporation and designated the T-59, it was standardized as the M-59 in 1953, and over 4000 vehicles were manufactured before production stopped in 1959. The M-59 is fully amphibious, being propelled in the water by its tracks. It has a hatched roof and a ramp at the rear.

M-75APC

Country of Origin: USA
Crew: 2 + 10

Engine: Continental AO-895-4, 6-cylinder, air-cooled petrol, developing 295hp at 2,000rpm
Performance: Speed 71km/h (44.1mph) road; range 185km (115mls); fuel 568 litres (125gals)
Armament: 1 × 12.7mm (0.5in) M2MG (1,800 rounds carried)
Armour: 25mm (1in) max
Dimensions: Length 5.193m (17ft); width 2.84m (9ft 3in); height 3.041m (9ft 11in) with MG, 2.775m (9ft 1in) inc cupola; ground clearance 0.457m (1ft 5in)

Weight: 18,828kg (41,422lb) loaded, 16,632kg (36,590lb) empty
Ground Pressure: 0.57kg/sq cm (8.15psi)
History: Now only in service with the Belgian Army, the M-5APC was quickly replaced in the United States Army with the M-59APC. Although some 1700-plus vehicles were built, the production being split between International Harvester and FMC Corporation in a three-year period ending in 1954, the M-75 suffered by not being amphibious.

M3A1
ARMOURED HALF-TRACK VEHICLE
Country of Origin: USA

Crew: 13

Engine: White Motor Company engine, developing 147bhp at 3,000rpm

Performance: Speed 73km/h (45mph) road; range 321km (199mls); fuel 227 litres (50gals)

Armament: 1 × 12.7mm (0.5in) M2HB MG, 1 × 7.62mm (0.3in) MG

Armour: 7–13mm (0.28 – 0.51in)

Dimensions: Length 6.337m (20ft 9in); width 2.22m (7ft 3in); height 2.692m (8ft 9in); ground clearance 0.28m (11ft)

Weight: 9.298kg (20,456lb) loaded, 6,940kg (15,268lb) empty

History: This famous vehicle was manufactured in great numbers by the Autocar Company, the Diamond T Motor Company and the White Motor Company during World War II. Still widely used in South America, the Middle East and Japan, variants included the M2 half-track car, M4A1 81mm (3.19in) mortar carrier, the M9A1 carrier, and the M16 multiple-gun motor carriage.

LVTP7
ARMOURED AMPHIBIOUS ASSAULT VEHICLE

Country of Origin: USA
Crew: 3+25

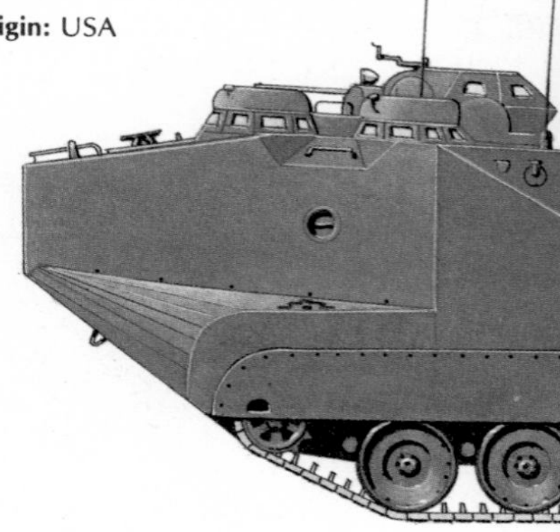

Engine: Detroit-Diesel model 8V53T, 8-cylinder, developing 400hp at 2,800rpm
Performance: Speed 64.31km/h (40mph) road, 13.5km/h (8.39mph) water; range 482km (300mls) road; fuel 681 litres (150gals)
Armament: 1 × 12.7mm (0.5in) M85 MG, elevation +60°, depression -10° (1,000 rounds carried)
Armour: 10mm-45mm – (90.4in-1.78in)
Dimensions: Length 7.94m (26ft); width 3.27m (10ft 8in); height 3.26m (10ft 8in) overall, 3.12m (10ft 2in)

turret; ground clearance 0.406m (1ft 3in)
Weight: 22,838kg (50,244lb) loaded, 17,441kg (38,370lb) empty
Ground Pressure; 0.57kg/sq in (8.15psi)
History: Originally designated LVTPX12, the first of 15

prototypes appeared in 1967. Developed by FMC, production vehicles were delivered to the USMC in 1971, and almost 1000 were built before production stopped in 1974. Variants include the upgraded LVTP7 with a Cummins VT904/400 engine, passive night-vision equipment, smoke-generating capability and an automatic fire detection and suppression system, designated LVTP7A1; and the LVTC7 tracked landing vehicle designated the Command Model 7, which has shelter-erecting capabilities and a crew of 13.

Index